FAMOUS REGIMENTS

The Northamptonshire Regiment

FAMOUS REGIMENTS

Edited by
Lt-General Sir Brian Horrocks

The
Northamptonshire
Regiment

(The 48th/58th Regiment of Foot)

by

Michael Barthorp

Leo Cooper Ltd, London

First published in Great Britain, 1974
by Leo Cooper Ltd
196 Shaftesbury Avenue
London WC2H 8JL

Copyright © 1974 by Michael Barthorp
Introduction Copyright © 1974
by Lt-General Sir Brian Horrocks

ISBN 0 85052 146 7

Printed in Great Britain by
Hazell Watson & Viney Ltd,
Aylesbury, Bucks

To J. A. F. B.
and to the memory of A. H. B. and M. A. R. B.
and of all ranks of
The Northamptonshire Regiment
who have fallen
in the service of
their country.

Contents

Illustrations

Plate No 2 is reproduced by gracious permission of Her Majesty the Queen. The author and publishers are also grateful to the following for permission to reproduce copyright pictures in this book:

The National Army Museum, Nos. 1, 5, 6, 7 and 8
The Prince Consort's Library, No. 3
The Alexander Turnbull Library, No. 10
R. G. Harris Esq., Nos. 11, 12 and 13
Trustees of the British Museum, Nos. 14 and 15
Imperial War Museum, Nos. 23 and 24

Preface

The Official History of the Northamptonshire Regiment is contained in four volumes: 1741–1934; 1914–1918; 1934–1948; and 1948–1960. It is the aim of this work to recall the deeds of the Regiment within one volume and, by so doing, to provide a concise and, it is to be hoped, readable record of a Regiment that no longer features in the Army List. It is not intended to be a précis of other men's work but rather a complementary volume to the existing histories. To achieve this, as much use as possible has been made of material and illustrations that do not figure in the official volumes. This history does not attempt to set down a detailed record of all the Regiment's activities in peace and war but to recreate the atmosphere of the major events of its past, particularly by using, in their original syntax and spelling, the words of those who experienced them.

Inevitably in a short history, the Regiment's part in the two World Wars cannot be covered in detail. This is fully dealt with in two volumes of the Official History, and so it has seemed more in keeping with the complementary nature of the book, to give here a summary of the Regimental effort in these two huge conflicts. This has also allowed due credit to be given to the part played by the Territorial and Service battalions, which would not otherwise have been possible.

In a work of this scope, much fact and many characters have had to be omitted. Some matters which are included, the Regiment would perhaps prefer to forget, but every regiment has had its vicissitudes and a history which disclosed only a shining tale of unparalleled glory would surely arouse scepticism in the mind of even its most devoted partisan.

In the last decade or so the British Infantry has suffered much change. The old regiments are in danger of being forgotten by the public, and the new ones are not fully understood. It is hoped that this book, like others in its series, will help to save one of the old County Regiments from oblivion.

Acknowledgements

I am grateful to the following members of the Regiment, who in various ways have given much help and advice: the late Major-General G. St. G. Robinson, CB, DSO, MC, the late Brigadier R. E. Osborne-Smith, DSO, OBE, Colonel G. V. Martin, MC, Lieutenant-Colonel F. R. Wilford, Major J. A. F. Barthorp, Major E. P. Kelly, DCM, and particularly Colonel J. B. Akehurst, who has read my manuscript and made many helpful suggestions. I owe a special debt to Major D. Baxter, who first suggested I undertake this task and who, with his staff at Northampton, has given great assistance.

I have received much help with the illustrations from several sources, notably from Mr R. G. Harris, of Southsea, Miss Allom of the National Army Museum and Lieutenant-Colonel R. E. J. Gerrard-Wright, OBE, Royal Anglian Regiment.

Extracts from the diary of Alec Whisker, 58th Foot, are reproduced by permission of the Auckland Institute and Museum, New Zealand, and those from the diary of Major Cyprian Bridge and the Reminiscences of Robert Hattaway, both of the 58th, by permission of the Alexander Turnbull Library, Wellington, New Zealand. I am grateful to the staffs of both these institutions for their help, and to Mr Peter Coates of The New Zealand Broadcasting Corporation for his kindness in directing my attention to these old records of the 58th.

Introduction

By Lt-General Sir Brian Horrocks

I should like to start by congratulating Major Barthorp on producing one of the best Regimental Histories it has been my privilege to read in this Series. Thanks to the immense amount of research he has undertaken, each battle is described through the eyes of someone who was actually present.

The 48th was formed on 3rd January, 1741 by Col the Hon James Cholmondley, an officer of the Life Guards, whose descendant was the Lord Great Chamberlain in the House of Lords when I was Black Rod. Their first action, against the Jacobites at Falkirk in Scotland on 17 January, 1746, was a disaster. Only two regiments kept their ground against the violent attacks of the Highlanders and emerged with their reputations unsullied. The 48th was one. Their first victorious battle was under the overall command of the Duke of Cumberland at Culloden Moor, on 16 April, 1746.

It is an interesting fact that, as early as 1755, when fighting against the French and Indians in Canada, the British troops (including the 48th Foot), who were forced to advance shoulder to shoulder in the open, already realised the importance of extended order and the use of cover. In the words of an eye witness: 'The soldiers then insisted, much, to be allowed to take to the trees, which the General denied, and stormed much, calling them cowards, and even went as far as to strike them with his own sword for attempting the trees.'

As usual the instinct of the men who actually do the fighting was right, but it was not until 1881 that our rigid tactical training was altered in the slightest detail.

Alarmed by the British Army's lack of success in Canada, the Government decided to increase the size of the Army and one of these new Regiments was the 58th Foot, which in 1757 was sent to America, where at Halifax it joined the 48th: the two future partners of the Northamptonshire Regiment thus met for the first time. During the subsequent conquest of Canada, the two Regiments were lucky enough to serve under General Wolfe, and played a prominent part in the capture of Quebec, (once more described by an eye witness), when their General died in the hour of victory. This is the first time I have ever read an intelligible account of the fighting in Canada.

During the eighteenth and nineteenth centuries the pattern for all the Infantry consisted of Major Wars to maintain the balance of power in Europe; operations in Ireland; training in the United Kingdom – where they were often used to keep order; and, by far the most unpleasant task of all, peace keeping operations involving long periods abroad. Of these, the most deadly was the West Indies. The 58th only served there twice, but as always suffered severely from the scourge of Yellow Fever. The 48th had the misfortune to spend 14 years in these dreaded isles. Four times between 1760 and 1800 they set sail to the West Indies as a complete battalion, and four times a pitiable remnant returned, their ranks decimated by disease.

Regimental affiliations to particular countries started towards the end of the eighteenth century, when the 48th was linked with Northamptonshire and the 58th with the neighbouring county of Rutland.

There is nothing I can add to Major Barthorp's excellent account of the various campaigns and battles in which the regiments took part during these years so I shall pass straight to the war against the Maoris in New Zealand in which the 58th took part. The Maoris were, and still are – as I can vouch from my own experience during the last war – first class fighters. In Sgt Robert Hathaway's account of the hardships endured by the troops, he describes how

'The parade-ground gloss soon disappeared – the Grenadier and Light Companies were a magnificent body of men . . . but their garments were ragged, tattered and torn . . . many without boots . . . their pants of many colours, but their morale remained high, and they were thoroughly disciplined and trained.'

He might well have been describing the appearance of the 8th Army at the end of the Desert Campaign. I can still remember the astonishment of the fresh, fully-equipped, and spick and span Americans, when they first saw Monty's Veteran Army, none of whom were wearing uniform, and whose bivouac areas looked like gipsy encampments, with those precious tin cans used for 'brew-ups', hanging round the tanks and vehicles.

But to return to New Zealand; despite the hardships, poor leadership (in some but not all cases) and defective commissariat, the Maori resistance was eventually worn down, and the campaign ended in July, 1847. When the 58th returned home eleven years later, a thousand men had taken their discharge, and remained on in New Zealand as colonists.

It would be wrong to pass over too lightly the battle of Laing's Nek, on 28 January, 1881, when the 58th formed part of a force sent to deal with the Boer Rebellion, on the Transvaal border. As the 58th advanced up the steep slope to attack the strongly-held entrenched position occupied by the Boers,

ite army returned from its foray into England and started to
ge Stirling. To raise this siege, General Hawley marched to
rk with a force of twelve battalions and three regiments of
oons. On 17 January, 1746, the two armies confronted one
her and the 48th's first action was about to begin.

he weather, which 'chang'd from fair to verry Rainy and windy',
Hawley's initial manoeuvres cannot have filled the men with
h confidence. Cholmondley, who had given up command of the
in 1742 and was now a Brigadier, wrote:

'About one, information was received that the rebel army was
arching towards us, and the army was immediately ordered to
and to arms in front of the encampment. The two lines of infantry
ere ordered to face to the left, and in this position we marched
hem to the left near a mile and a half, but as we had hollow roads
nd very uneven ground to pass over we were in great confusion.
Here we formed again, in my opinion a very good situation, but we
were no sooner formed than ordered again to take ground to our
eft, and as we marched all the way up hill and over uneven ground
he men were greatly blown.'

Eventually Hawley was satisfied and the Royal army was again
awn up in two lines with the cavalry on the left. The 48th were
cond from the right in the front line between the 1st and 14th Foot;
large ditch lay between them and the enemy. The rain by now had
rned to sleet which blew with gale force into the faces of the Royal
oops. The battle opened disastrously. The dragoons went forward
nto the storm and were met with a volley from the Highlanders
wo of the regiments turned about and galloped off in disorder
ollowed shortly afterwards by the third. Then the Highlanders
harged. The Royal infantry, blinded by the sleet, appalled by t
udden disaster to the cavalry, and with their powder affected
he storm, received the charge with a feeble volley. The Highland
returned the fire and then threw themselves on the English left, wh
broke under the broadswords. Panic and chaos spread rapidly a
as one eyewitness noted with masterly understatement, 'some f
Regts did not perform wonders'. Only two regiments failed to br
as Cholmondley described:

'In column of companies they made a brave show, in their white
helmets and red coats. For the last time a British battalion was
going into action in the old style – officers mounted, men in close
order, colours flying, yet the enemy were skilled marksmen, armed
with modern rifles.'

The result was complete disaster although there was no lack of
courage; Lt Hill earned the Regiment's first VC for staying behind on
his own to rescue two wounded men, but the British force was routed.

At Majuba, another and better known defeat, the Boers drove the
British force off the hills which they had occupied in darkness, by
skilled fire and movement, a tactic then unknown in the British Army.
This was the last battle fought by the 58th as an individual regiment.
The Cardwell Reforms of 1872 linked the 48th and 58th, which subsequently in 1881 became the Northamptonshire Regiment, with their
own 1st, 2nd, 3rd and 4th Volunteer Battalion.

The 48th did four tours of duty in India; in the last one they distinguished themselves during a particularly hard campaign in 1897,
against the Tribesmen on the North West Frontier.

I mention these last two operations, the 58th in South Africa and
the 48th on the North West Frontier, because in each case the Regiment had been opposed by first class marksmen who made the best
possible use of the cover available. This lesson was taken to heart. Very
soon the Northamptons earned the reputation of being one of the best
– if not the best – shooting regiment in the British Army, and this stood
them in very good stead during the Boer War. The fact that the 58th
advanced in extended order, and were highly skilled marksmen, saved
them the appalling casualties which were suffered by the Guards during the relief of Kimberley.

I shall now pass straight on to the Second World War, in which six
Battalions served. The 2nd and 5th took part in the withdrawal to
Dunkirk and, for a very short time, the 5th Battalion came under my
command in the Beachhead. No Brigadier, however, has ever had less
influence on the troops under his command. Twenty-four hours after
being appointed command of the 11th Brigade, we were ordered to
evacuate our position on the defensive perimeter by platoon, and then
move down to La Panne, where ships were available to evacuate us to
the United Kingdom. Alas, the tide was out and the ships were beyond
reach. Moreover, the beach at La Panne was being shelled by the Germans, but, like the other two battalions in the Brigade, the troops
showed no panic whatever, and I last saw them marching to Dunkirk,
where they embarked for the UK, reaching Dover on 1 June. I arrived
in the United Kingdom to find that my temporary appointment as
Brigadier had been cancelled.

xvi

I never met the 58th which, by the time I reached the desert, were in India. Meanwhile, the 5th Battalion had joined one of the most famous fighting formations of the last war – the 78th Battle Axe Division, which was part of the 1st Army which landed in Algiers in November, 1942. During the next 6 months they took part in some of the hardest fighting of the war, during which the battalion had 740 casualties. Although this Division did not take part in the actual capture of Tunis it was largely their battles just north of Long Stop Hill that made the final attack up the Madjerba Valley possible and they entered the town the day after its capture.

The 58th and 5th now came under command of the 8th Army for the operations in Sicily. The Battle Axe Division always felt that they had never been given the credit which was their due for their hard fighting during the final stages of the war in North Africa while the 8th Army's operations had captured all the headlines. Consequently, there was no love lost between the two. So, when they landed in Sicily, their vehicles were inscribed in large letters with the words, 'We have no connection with the 8th Army'!

A note of bitterness creeps into this History in the last Chapter, when the author describes the post-war amalgamations which, by 1960, resulted in the Northamptons losing their identity and eventually becoming part of the Royal Anglian Regiment. I can well understand his feelings but I would remind him, however, that the Army has always suffered severe reduction at the end of every war. We are not, thank goodness, a military nation and our system of democracy demands in peace time economies in the defence budget.

I shall end with that famous French quotation:

'Plus ça change, plus c'est la même chose.'

Should another major war erupt, the County of Northampton will, I am sure, rally to the Flag as swiftly and eagerly as it always has done in the past.

CHAPTER

Formation and Early

A CONTEMPORARY author, writing on referred to what he calls, 'the absol ments of the Line'.* By this term Highlanders, Fusiliers or Light Infantry, but county regiments of England; regiments endured in victory and defeat, in camp an years; receiving scant acclaim or recognitio endeavours the nation could not have survived Northamptonshire Regiment, a union of the ol ments of Foot, with 219 years' service to the C able record in most of England's wars, fro Africa, from Canada to New Zealand.

It was war that brought the 48th Foot to b Austrian Succession required the formation of r 3 January, 1741, Col the Hon James Cholmond Life Guards, was ordered to raise 'a Regiment o of ten Companeys of three Sergeants, three Co mers and seventy effective private men in each Commission Officers'. The Regiment, which form precedence initially as the 59th of the Line, but a of 48th in 1748. The uniform was red with buff fac

Three years after its formation, the Regim Flanders to join the army of the Duke of Cumber August, 1745 the news of the advance into Engl Pretender with his army of Highlanders caused return troops to meet this threat. By the time the 4 land, the Pretender had moved south and the Reg was to occupy Edinburgh. They were still in garriso

* Simon Raven, 'Perish by the Sword' in *Boys will* Blond, 1963, page 25.

N.R.—2

'In column of companies they made a brave show, in their white helmets and red coats. For the last time a British battalion was going into action in the old style – officers mounted, men in close order, colours flying, yet the enemy were skilled marksmen, armed with modern rifles.'

The result was complete disaster although there was no lack of courage; Lt Hill earned the Regiment's first VC for staying behind on his own to rescue two wounded men, but the British force was routed.

At Majuba, another and better known defeat, the Boers drove the British force off the hills which they had occupied in darkness, by skilled fire and movement, a tactic then unknown in the British Army. This was the last battle fought by the 58th as an individual regiment. The Cardwell Reforms of 1872 linked the 48th and 58th, which subsequently in 1881 became the Northamptonshire Regiment, with their own 1st, 2nd, 3rd and 4th Volunteer Battalion.

The 48th did four tours of duty in India; in the last one they distinguished themselves during a particularly hard campaign in 1897, against the Tribesmen on the North West Frontier.

I mention these last two operations, the 58th in South Africa and the 48th on the North West Frontier, because in each case the Regiment had been opposed by first class marksmen who made the best possible use of the cover available. This lesson was taken to heart. Very soon the Northamptons earned the reputation of being one of the best – if not *the* best – shooting regiment in the British Army, and this stood them in very good stead during the Boer War. The fact that the 58th advanced in extended order, and were highly skilled marksmen, saved them the appalling casualties which were suffered by the Guards during the relief of Kimberley.

I shall now pass straight on to the Second World War, in which six Battalions served. The 2nd and 5th took part in the withdrawal to Dunkirk and, for a very short time, the 5th Battalion came under my command in the Beachhead. No Brigadier, however, has ever had less influence on the troops under his command. Twenty-four hours after being appointed command of the 11th Brigade, we were ordered to evacuate our position on the defensive perimeter by platoon, and then move down to La Panne, where ships were available to evacuate us to the United Kingdom. Alas, the tide was out and the ships were beyond reach. Moreover, the beach at La Panne was being shelled by the Germans, but, like the other two battalions in the Brigade, the troops showed no panic whatever, and I last saw them marching to Dunkirk, where they embarked for the UK, reaching Dover on 1 June. I arrived in the United Kingdom to find that my temporary appointment as Brigadier had been cancelled.

I never met the 58th which, by the time I reached the desert, were in India. Meanwhile, the 5th Battalion had joined one of the most famous fighting formations of the last war – the 78th Battle Axe Division, which was part of the 1st Army which landed in Algiers in November, 1942. During the next 6 months they took part in some of the hardest fighting of the war, during which the battalion had 740 casualties. Although this Division did not take part in the actual capture of Tunis it was largely their battles just north of Long Stop Hill that made the final attack up the Madjerba Valley possible and they entered the town the day after its capture.

The 58th and 5th now came under command of the 8th Army for the operations in Sicily. The Battle Axe Division always felt that they had never been given the credit which was their due for their hard fighting during the final stages of the war in North Africa while the 8th Army's operations had captured all the headlines. Consequently, there was no love lost between the two. So, when they landed in Sicily, their vehicles were inscribed in large letters with the words, 'We have no connection with the 8th Army'!

A note of bitterness creeps into this History in the last Chapter, when the author describes the post-war amalgamations which, by 1960, resulted in the Northamptons losing their identity and eventually becoming part of the Royal Anglian Regiment. I can well understand his feelings but I would remind him, however, that the Army has always suffered severe reduction at the end of every war. We are not, thank goodness, a military nation and our system of democracy demands in peace time economies in the defence budget.

I shall end with that famous French quotation:

'Plus ça change, plus c'est la même chose.'

Should another major war erupt, the County of Northampton will, I am sure, rally to the Flag as swiftly and eagerly as it always has done in the past.

CHAPTER 1

Formation and Early Campaigns

A CONTEMPORARY author, writing on the British Army, has referred to what he calls, 'the absolutely plain run of Regiments of the Line'.* By this term he defines, not Rifles or Highlanders, Fusiliers or Light Infantry, but the unsung, unassuming county regiments of England; regiments which have fought and endured in victory and defeat, in camp and in garrison, for 300 years; receiving scant acclaim or recognition, but without whose endeavours the nation could not have survived. Such a corps was The Northamptonshire Regiment, a union of the old 48th and 58th Regiments of Foot, with 219 years' service to the Crown and a commendable record in most of England's wars, from Scotland to South Africa, from Canada to New Zealand.

It was war that brought the 48th Foot to birth. The War of the Austrian Succession required the formation of new regiments and on 3 January, 1741, Col the Hon James Cholmondley, an officer of the Life Guards, was ordered to raise 'a Regiment of Foot . . . to consist of ten Companeys of three Sergeants, three Corporals, two Drummers and seventy effective private men in each Company, besides Commission Officers'. The Regiment, which formed at Norwich, took precedence initially as the 59th of the Line, but assumed the position of 48th in 1748. The uniform was red with buff facings.

Three years after its formation, the Regiment embarked for Flanders to join the army of the Duke of Cumberland. However, in August, 1745 the news of the advance into England of the Young Pretender with his army of Highlanders caused Cumberland to return troops to meet this threat. By the time the 48th reached Scotland, the Pretender had moved south and the Regiment's first task was to occupy Edinburgh. They were still in garrison there when the

---

* Simon Raven, 'Perish by the Sword' in *Boys will be Boys*, Anthony Blond, 1963, page 25.

N.R.—2

Jacobite army returned from its foray into England and started to besiege Stirling. To raise this siege, General Hawley marched to Falkirk with a force of twelve battalions and three regiments of dragoons. On 17 January, 1746, the two armies confronted one another and the 48th's first action was about to begin.

The weather, which 'chang'd from fair to verry Rainy and windy', and Hawley's initial manoeuvres cannot have filled the men with much confidence. Cholmondley, who had given up command of the 48th in 1742 and was now a Brigadier, wrote:

'About one, information was received that the rebel army was marching towards us, and the army was immediately ordered to stand to arms in front of the encampment. The two lines of infantry were ordered to face to the left, and in this position we marched them to the left near a mile and a half, but as we had hollow roads and very uneven ground to pass over we were in great confusion. Here we formed again, in my opinion a very good situation, but we were no sooner formed than ordered again to take ground to our left, and as we marched all the way up hill and over uneven ground the men were greatly blown.'

Eventually Hawley was satisfied and the Royal army was again drawn up in two lines with the cavalry on the left. The 48th were second from the right in the front line between the 1st and 14th Foot; a large ditch lay between them and the enemy. The rain by now had turned to sleet which blew with gale force into the faces of the Royal troops. The battle opened disastrously. The dragoons went forward into the storm and were met with a volley from the Highlanders; two of the regiments turned about and galloped off in disorder, followed shortly afterwards by the third. Then the Highlanders charged. The Royal infantry, blinded by the sleet, appalled by the sudden disaster to the cavalry, and with their powder affected by the storm, received the charge with a feeble volley. The Highlanders returned the fire and then threw themselves on the English left, which broke under the broadswords. Panic and chaos spread rapidly and, as one eyewitness noted with masterly understatement, 'some foot Regts did not perform wonders'. Only two regiments failed to break, as Cholmondley described:

'Barrell's Regiment * kept their ground, and I got my late regiment (48th) to form on their right. In this situation we kept our ground and, with the assistance of the officers (who deserve the greatest praise for the spirit they showed), I got the men to be quite cool, as cool as I ever saw men at exercise, and when the rebels were down upon us, we not only repulsed them but advanced and put them to flight.'

While these two battalions fought on, attempts were made to rally the rest of the army. This was eventually accomplished but as night was coming on and the storm continued unabated, General Hawley ordered a retirement. The 48th's first battle thus ended in defeat, although they had emerged from the ordeal with credit. As another eyewitness put it, 'Ligoniers (48th) and Barrels distinguish'd themselves . . . the behaviour of Ligoniers young Regt. is greatly owing to the activity and care of the Lt-Coll'.

Cumberland now arrived to command the army, which again advanced to bring the Jacobites to battle. However, it was not until April that contact was made with the Pretender. On 16 April Cumberland moved forward to give battle on Culloden Moor outside Inverness. The approach of Cumberland's troops was thus described by a soldier of the 1st Foot, Alexander Taylor:

'The Army was formed by Five, and we were on our long March a quarter after Five. It was a very cold rainy Morning, and nothing to buy to comfort us: But we had the Ammunition-loaf, thank God; but not a Dram of Brandy or Spirits, nor nothing but the Loaf and Water. We had also great Difficulty in keeping the Locks of our Firelocks dry: which was absolutely necessary; for the Day was stormy, and the Rain was violent. We marched but four Miles till we was alarmed by their Out-parties, and drew up in Order of Battle, and marched that Way for two Miles, with our Arms secured, and Bayonets fixed, (a very uneasy Way of marching); and then getting Intelligence where they were, we took our long March again, and marched so, till we came in sight of their Lines.'

The army halted about 500 yards from the enemy and prepared for action.

* At this period regiments were usually known by their Colonel's name. Barrell's was the 4th Foot. Although at Falkirk the 48th were known as Ligonier's, by April they had become Conway's.

According to a contemporary diagram of the battle, the 48th, who went into action 396 strong, were placed second from the left in the second line with the 8th Foot on their left and the 25th and 20th on their right; but just before the battle began, the 8th were moved forward to a position at right angles to the left battalion of the first line, the 4th Foot.* Taylor's description continues:

> 'The Battle begun by Cannonading, and continued for half an Hour or more with great Guns. But our Gunners galling their Lines, they betook them to their small Arms, Sword and Pistol, and came running upon our Front-Line like Troops of hungry Wolves, and fought with Intrepidity.'

The main fury of their charge fell upon the 4th Foot and the 37th on the British left. Though forced back and disordered by the Highland charge, these two battalions, assisted by the flanking fire of the 8th, fought savagely to contain the onslaught. Then the 25th and 20th moved forward in support, while the 48th advanced and wheeled right to enclose the Highlanders between the 4th and 25th. Together these six battalions completely decimated the main Jacobite attack. Cumberland then ordered his cavalry forward and the Pretender's forces were driven from the field.

For a young regiment in its first battles, the 48th had achieved distinction, for which much was due to its commanders. The competence of their Lieutenant-Colonel, George Stanhope – by no means a *sine qua non* in those days – has already been mentioned; they were also fortunate in their early Colonels, James Cholmondley, Francis Ligonier and Henry Conway. At this period many Colonels of regiments were 'absentee landlords' but these three were true soldiers and their stewardship of the Regiment in its early years provided a solid foundation on which their successors could build.

The reputation gained in the '45 Rebellion was soon to be put to the test. After returning to Flanders, the Regiment was engaged at the Battle of Laffelt in 1747 against the French under Marshal Saxe.

---

* Some later accounts, including Russell Gurney's history of the Regiment, place the 48th in the centre of the second line to the right of the 20th. This seems unlikely as in the order of march onto the battlefield, they were between the 8th and the 25th.

After the French captured the village of Val, the 48th, with the 8th and 19th, were ordered to retake it. The village was stormed and won but, as the French deployed their powerful reserves, the British were forced out. Again they charged but again the French counter-attacked. The Duke of Cumberland wrote: 'They rallied and charged into the village four or five times each; the French but once as they could not be rallied but were always replaced by fresh troops ... the infantry behaved so well, one and all, that he could not commend any one regiment without doing injustice to the rest.' The 48th's casualties were severe; Col Conway was taken prisoner, Lt-Col Stanhope was wounded and more men were killed than in any other regiment. However, it was their last battle of the campaign and when peace was signed in 1748, they sailed for Ireland.

Seven years' garrison duty in Ireland cannot have been the best preparation for their next campaign; a campaign which was to introduce the 48th to a type of warfare for which neither they nor any regiment of British infantry had ever been trained. Equally unpromising for the future was the appointment, in 1752, of Henry Dunbar as Colonel, an officer whose supine conduct in the forthcoming campaign was to result in his premature departure from the Regiment. Furthermore the Regiment, when ordered on active service, was under strength and had to be made up by men drafted unwillingly from other regiments, plus whatever recruits could be enlisted at its destination. Such was the inauspicious start to Maj-Gen Braddock's expedition to capture Fort Duquesne from the French forces in Canada.

The only two Regular regiments designated for the expedition, the 44th and 48th, arrived in America late in March, 1755. For the next two months, Braddock struggled with avaricious and uncooperative colonists to equip his force. By 10 June all was ready and the march began. The force consisted of 1,400 men of the 44th and 48th, 100 gunners, thirty seamen from HMS *Garland*, 450 Virginian Militia and fifty Indians. Behind the scouts and advance guard, 300 axemen hacked a path through the undergrowth; along the track thus made rolled the long column of guns, waggons and pack-horses with the Regulars and Militia marching alongside; through the forest on either flank, small groups of infantry forced their way to guard against surprise. Progress was painfully slow and the damp of the forest soon caused fever among the men. At night around the camp fires, the militiamen curdled the blood of the Regulars with terrifying tales of Indian ferocity. By 18 June, when the force had only covered thirty miles, Braddock received reports of French reinforcements approaching the Fort. Leaving a third of his force and the heavier baggage under Col Dunbar to follow on at their best speed, he pressed on with the remainder.

On 9 July, when only six miles from the Fort and as the troops moved out into more open country, the advance guard was suddenly fired on. What ensued is described in the Journal of the seamen from the *Garland*:

> 'The Advance Party was now about a quarter of a mile before the main body, the rear of which was just over the river (Monongahela) when the front was attacked. The two Grenadier Companies formed the 2 flank advance Picquets, 2 Companies of Carpenters cutting the Roads, and the rest covering them. The first fire our men received was in front, and on the flank of the flank Picquets, which in a few minutes nearly cut off the most part of the Grenadiers and the Company of Carpenters. As soon as the General with the main body heard the Front was attacked, they hastened to succour them, but found the Remains retreating. Immediately the General ordered the cannon to draw up and the Batallion to form. By this time the Enemy began to fire on the Main Body, which faced to the right and left and returned it, and the Cannon began to play, but could not see at what, (for our men were formed in the open road they had just cut, and the Enemy kept the Trees in front and on the flanks).'

Another eyewitness wrote:

> 'The Enemy kept behind Trees and Loggs of Wood, and cut down our Troops as fast as they cou'd advance. The Soldiers then insisted much to be allowed to take to the Trees, which the General denied and stormed much, calling them Cowards, and even went so far as to strike them with his own Sword for attempting the Trees.'

The seamen's Journal continues:

> 'On the right they had possession of a hill which we could never get possession of, though our Officers made many attempts to do it:

but if the Officers dropped which was generally the case, or the Enemy gave a platoon of ours advancing up the hill a smart fire, they retreated down again. As numbers of our Officers declared they never saw above 4 of the enemy at a time the whole day, it struck a panic through our men to see numbers daily falling by them, and even their comrades scalped in their sight. As soon as the General saw this was the case, he ordered that our men should divide into small parties and endeavour to surround the Enemy, but by this time the greatest part of the Officers were either killed or wounded, and in short the Soldiers deaf to the commands of those few that were left alive.'

Braddock was mortally wounded; appeals for help to Dunbar went unanswered. The wretched Regulars, trained to fight shoulder to shoulder against a visible enemy, fired off their last rounds against the invisible Indians and Frenchmen who surrounded them. For two and a half hours they had stood against this terrible musketry, but now, their officers down, they broke and the remnants of Braddock's column fled back the way they had come. The losses had been punishing; 700 were killed or wounded; eighteen of the 48th's officers were casualties and in the Grenadier Company only nine men escaped out of seventy-nine.

The recriminations that followed Braddock's defeat were unedifying; a cynic summed it up in the *Gentleman's Magazine* for September, 1755:

> 'Ah! Braddock, why did you persuade
> To stand and fight each recreant blade
>   That left thee in the wood?
> They knew that those who run away,
> Might live to fight another day,
>   But all must die that stood.'

At least the 48th now knew what to expect of American warfare. There followed a period of retraining and recruitment, many colonists being enlisted. Their dress, however, caused the Commander-in-Chief in North America, Lord Loudon, some concern for, as he observed to the Duke of Cumberland, their coats were 'so thorough worn, that they are really like Cobwebs'. Nevertheless, by

1756 he felt able to speak of the Regiment as one of his best corps. A reverse has often proved salutary to British arms.

One side effect of this disastrous campaign was to prove of significance to the future Northamptonshire Regiment. Alarmed by events in America, the Government decided to increase the Army by eleven new battalions. The junior of these, raised as the 60th Foot in December, 1755, by Col Robert Anstruther, was two years later to become the 58th. A nucleus for the new Regiment came from the 12th and 37th Foot, who each provided a draft of sixteen NCOs, four Drummers and forty 'of the best Private men'. The uniform was 'red, faced and lapelled with black, lined buff'; the black facings being derived from the Anstruther coat of arms.

In 1757 the 58th was sent to America. In July of that year, at Halifax, the two future halves of The Northamptonshire Regiment met for the first time. Together they would avenge, in the forthcoming campaign, the defeat of the 48th on the banks of the Monongahela.

CHAPTER 2

The Conquest of Canada

ONE officer of the 58th was a man, 'six feet in height, of coarse mould and exceedingly dark', who was to make a brilliant name for himself in Canada and was to achieve high rank and some distinction in his later career. This was the Hon William Howe, the third son of Viscount Howe. His early commissioned service had been in the 20th Foot, as a brother officer of James Wolfe, and, at the age of twenty-six, he had been appointed Major of the 58th on its formation. On 17 December, 1757, he was promoted Lieutenant-Colonel of the Regiment. By a strange augury, his new command was to receive its baptism of fire alongside its future partner, the 48th, in the first battle for Canada.

The objective was Louisburg, the great French fortress on Cape Breton Island, looking out over the Atlantic and guarding the mouth of the St Lawrence River, the life-line of French Canada. The British force of fourteen battalions arrived off Louisburg on 3 June, 1758. The weather was vile, 'wind very hard at South . . . the fog excessive thick and wet', and a landing was out of the question. For four days the storms and fogs continued, but on the 8th, under cover of the fleet, the troops were rowed towards the French entrenchments along the shore.

The right wing was commanded by Brig Whitmore, described by his fellow brigadier, James Wolfe, as 'a poor, old, sleepy man', and consisted of two brigades, one commanded by Lt-Col Ralph Burton of the 48th, and the other formed of the 48th, 58th and 17th Foot. In the centre was Brig Lawrence's wing and on the left, leading the assault, was Wolfe's force: the Grenadiers and Light Infantry of all the regiments, supported by the 78th Highlanders. As the boats ploughed through the foaming surf, the French guns, concealed behind an abattis of felled trees, opened fire at close range. A storm of grapeshot shattered the leading craft and troops were hurled into the heavy sea. Wolfe signalled a retirement but three boats, containing one hundred Light Infantry under Lts Thomas Hopkins, 48th, John Grant, 58th, and an officer of the 60th discovered a small beach concealed behind rocks. They stormed ashore and formed a beach-head. Wolfe directed the remainder of his force to this point and rapidly carried the enemy position.

The rest of the Army were landed and the fortress invested. Throughout June and into July, in wind, rain and fog, the British siege works crept closer to the walls. On 22 July the British batteries opened fire and set part of the fortress alight. For four days the bombardment was continued until on the 27th the French garrison surrendered. The gateway to the St Lawrence was open. Leaving a garrison at Louisburg, the regiments sailed away to their winter quarters. In their first action the 58th must have made their mark, for Wolfe wrote in a letter: 'Our old comrade Howe is at the head of the best trained battalion in all America'.

Wolfe too had made his mark, for the following year he was appointed to command the force of eleven battalions earmarked for the capture of Quebec. Having successfully navigated the St Lawrence, Wolfe arrived off the Isle of Orleans, four miles downstream from the city, and disembarked his troops on 26 June, 1759.

Apart from the fortifications of the town itself, situated on its headland on the north bank of the river, the main French defences lay between the city and the River Montmorency, which fell through a gorge into the St Lawrence seven miles to the east. Opposite the town, where the river flowed through a channel three-quarters of a mile wide, the Point Levis stood on a headland on the south bank. Here Col Burton and the 48th were posted to protect a siege battery from the attacks of marauding Canadians and Indians; this was to remain the Regiment's main task throughout the campaign. The greater part of the Army, including the 58th, was moved on 8 July to the north bank, just east of the Montmorency, where artillery batteries were mounted to enfilade the French defences.

The 58th was now temporarily commanded by Maj James Agnew, for Howe had been given command of a battalion of Light Infantry, formed of men from all the regiments who were 'good marksmen, active marchers and alert spirited soldiers able to endure fatigue'. The Light Infantry had their clothing and equipment specially adapted for forest fighting and tomahawks were added to their armament. Howe soon began to learn his new trade on the banks of the Montmorency, reconnoitring, testing the French defences and fighting the Indian scouting parties with their own tactics.

Patrols were also despatched upstream beyond Quebec to obtain information, to deceive the French and to try and bring them to battle. On 18 July the Grenadier Company of the 48th formed part of a force under Col Carleton which landed twelve miles above the town and dealt speedily with some Indians and Canadians, taking some sixty prisoners, 'mostly women'. This patrol did have the effect of making Montcalm, the French commander, detach a small force to guard that line of approach but, in common with other feints, it did not induce the French to leave the shelter of their defences around the city.

Towards the end of July, Wolfe realized he would have to force the issue. He decided to attack the French position on the cliffs one mile upstream from the Montmorency gorge. Whilst Howe's Light Infantry demonstrated in the woods along the Montmorency to simulate an attack, a force, under Wolfe's personal command and spearheaded by the massed Grenadiers of the Army led by Burton of the 48th, was to attack the French entrenchments from the St Lawrence.

The afternoon of 31 July was hot, sultry and foreboded a storm. At about five pm, under cover of a cannonade from the fleet and the land batteries, Burton's Grenadiers charged ashore and hurled themselves at a French redoubt on the beach. Their dash and enthusiasm proved their undoing. Instead of forming and waiting for further orders, they rushed to the cliff-face and started to haul themselves up to attack the defences above. An NCO of the 58th, John Johnson,* described what happened: 'The very first fire they received from the enemy was so well directed, that it checked their

---

* Quarter-Master Sergeant Johnson was clerk of the 58th and kept a memorandum of events and a journal during the Quebec campaign. He re-wrote them in narrative form when he entered Chelsea Hospital in 1784. His book was published in 1788.

impetuosity; and which being so well maintained, that they were soon put into the utmost confusion and were obliged to take Shelter ... until the troops were all landed.' Wolfe ordered their withdrawal and those still alive scrambled down to form again on the beach but 'leaving behind them their Killed and Wounded, to the tender mercy of the Indian and Canadian Savages . . . who murdered and scalped all that lay in their way.' Although the rest of the assault troops were now in position, any further attack seemed hopeless. Johnson summed up Wolfe's predicament. 'Looking up and seeing no probability of Success; and being filled with horror at the barbarous cruelties committed before our eyes; as also the day began to wear away apace; and the men in great measure intimidated at the seeming improbability of Success; and standing all the while under the most severe fire from their entrenchments; and the tide coming in a great pace', there appeared no alternative but withdrawal. The threatened storm broke and under its cover the British re-embarked; Wolfe's first attempt at Quebec had failed utterly.

The following day Wolfe's orders informed the Grenadiers that: 'Such impetuous, irregular and unsoldierlike proceeding destroys all order, makes it impossible for their Commanders to form any disposition for an attack, and puts it out of the General's power to execute his plan.' In addition, if an engineer officer, Capt Hollandt, heard correctly, Wolfe was 'much out of humour with Colonel Burton', though this does not seem to have permanently affected Wolfe's regard for Burton.

Throughout August Wolfe wrestled with the problem of taking Quebec. Exhausted by work and anxiety, at odds with his brigadiers, his health, never good, now collapsed. However, his will drove him on and, following a Council of War, a new plan was decided upon. The chaplain of the 48th may have contributed to this plan. The Rev Michael Houdin, who had been a missionary to the Mohawk Indians, had an extensive knowledge of, and contacts in, Quebec and combined his priestly duties with those of an intelligence officer; he was thus well-qualified to suggest where an attack might succeed. The Army was to move away upstream some ten miles and threaten a landing there; then, whilst the rest of the fleet feinted against the

French defences between the Montmorency and the city, the Army would return by night, scale the cliffs at the Anse du Foulon below the Heights of Abraham and give battle outside the walls of Quebec. During the first week of September the Army was moved up-river well above Quebec; only the 48th and 3rd/60th remained guarding the battery at Point Levis. The plan was for the first six units to be ferried to the landing place in boats, which would then return to the transports to bring down the next wave, finally crossing over to the south bank to collect the 48th and 3rd/60th. The Light Infantry was to land first and the 58th was to be fifth in order of disembarkation. Leading the Light Infantry were twenty-four men of the 'forlorn hope'.* Howe had had some difficulty in finding these, as when volunteers were called for, only eight men stepped forward; those eight were therefore instructed to each select two other 'volunteers'.

At nine pm on the 12th, the leading troops filed into the boats. At about one-thirty the tide began to ebb and two lights appeared in the maintop-mast shrouds of HMS *Sutherland*. With this signal the boats began to slip silently down the river with muffled oars. Past the French sentries they floated, the challenges answered by a French-speaking Highlander, until the leading boat grounded on the beach at the Anse du Foulon. Immediately the 'forlorn hope', followed by Howe and the Light Infantry, started to scramble up the brush-covered 200-foot cliff. At the top a French picquet was rushed and the landing place secured.

The other battalions of the first wave were hurried up the path to the heights above. When the 58th were up, they were detached with Howe's Light Infantry under Brig Murray to silence a battery to the west of the landing place. Howe moved round to a flank, while in the dawn Murray and the 58th advanced stealthily through the pine trees. Before they could reach the battery however, an officer from Wolfe arrived with an order to return immediately. Murray turned them about but sent on the officer with a platoon of the 58th Grenadiers to overtake the Light Infantry. Before they could make contact, they came upon the battery. The French opened fire and

* 'Forlorn hope' – the party of men who led an attack or were first into a breach; so called because chances of survival were minimal.

the Grenadiers boldly charged the position, as Howe's men burst in from the flank. Having immobilized the battery, the whole party rejoined the main force.

Meanwhile Wolfe had been forming his line of battle. The 58th were positioned on the extreme left, except for the 15th Foot who were formed on their left rear to cover the flank. The 48th, the last to land, formed the only reserve. The line had hardly been formed when skirmishing developed on the left with Canadians and Indians approaching from the woods, and patrols from the 58th were sent forward to deal with these. At ten am the main action began. The story is familiar; the thin English line, only two ranks deep, three feet between files and forty yards between battalions; the steady advance of the French regulars and Canadian militia; at a range of thirty-five yards, what Fortescue called 'the most perfect volley ever fired on a battlefield'; the British charge and pursuit to the walls of Quebec; and Wolfe's death in the hour of victory. It was a great victory, and a source of pride to The Northamptonshire Regiment that both 48th and 58th were present.

It was a tradition in the Regiment that Wolfe was attended at his death by a surgeon and a Grenadier of the 48th. The claim for the surgeon can be substantiated by a letter from Ensign William Johnston of the 48th, who wrote: 'Mr Watson Surgeon to our Regt dressed Mr Wolfe's woonds'; his evidence is partially corroborated by Capt Hollandt who spoke of 'the Surgeon's Mate of that Regiment (48th), the only medical person who appeared, endeavoured to afford assistance'. The claim for the Grenadier is more flimsy, based as it was on the buff facings of the Grenadier in West's painting of Wolfe's death. Eyewitness accounts suggest that the Grenadier in question was most probably Volunteer James Henderson of the 22nd Foot (who also had buff facings), serving with the Louisburg Grenadiers.*

Quebec was won and now had to be held throughout the approaching winter. The town and its fortifications were badly damaged, the

* The Louisburg Grenadiers, who were present at Quebec, were formed from the Grenadier Companies of the 22nd, 40th and 45th Foot, part of the garrison of Louisburg.

weather would be severe and there was no hope of relief or rein-forcement until the spring. Brig Murray was appointed Governor, and Col Burton Lieutenant-Governor. Work was begun to repair the defences, to construct billets and to provide fuel for the winter. To give early warning of any French attack, outposts were estab-lished well beyond the town.

A detachment of thirteen men, under Sgt Carruthers of the 58th, found itself 'hemmed in an all Sides by such a number of Savages and their Retreat was intirely cut off'. Expecting to be attacked and killed 'in the most inhuman and barbarous manner . . . the sergeant immediately gave orders to put themselves in a posture of defence and made a disposition to attack a large body of those Cannibals'. On seeing Carruthers' determination, the Indians abandoned the attack; but then 'he found himself lost, and bewildered in the wood, not knowing whether he was going out or further into the wood'. However, he pressed on and though he encountered further parties of Indians and lost two men killed, he eventually reached safety; an escape ascribed by the devout QMS Johnson to the 'Good God, whose Gracious and Mercifull eye saw their distress'. Carruthers refused the commission Murray offered him but accepted thirty guineas and the post of Master-Gunner of Carlisle on his retirement.

Throughout the winter, conditions in Quebec worsened: intense cold, disease, shortage of food, no pay for the men and endless labour and vigilance. Everyone had to work at the defences; even officers were 'yoked in the Harness dragging up Cannon from the Lower Town, at work at the Batteries with the Barrow, Pick ax, and Spade'. Johnson remarks without irony that the soldiers felt 'anxious grief to see their Officers doing the common labour of the soldiers'.

In January, 1760, the French attacked by night across the ice-bound St Lawrence but were beaten off. In April they came on again with 9,000 men and Murray marched his troops out to give battle on the Heights of Abraham. He had barely 3,000 men and many of these were 'half-starved, scorbutic skeletons'. The line of ten weak battalions formed. Col Burton commanded the right brigade, which included his 48th and the 58th. Murray seems to have intended to throw up trenches and fight defensively but seeing the French line not fully deployed, he ordered an attack.

The light troops on either wing went forward, but, after initial success, began to be forced back by superior numbers and the British flanks were in danger. The whole French line now advanced and, as Johnson describes, 'by maintaining a Severe Musquetry upon us, we were obliged to give way. Their Indians and Canadians, like a hasty torrent, broke through the intervals and bore down all before them, and obliged us to a hasty and precipitate retreat'.

The French now besieged the garrison in earnest but they were too late. The thaw had set in, the St Lawrence was open and by 16 May three British ships lay off Quebec. At midnight on the 17th a British patrol found the French trenches deserted. Quebec was secure at last.

There now only remained the capture of Montreal, which was accomplished on 8 September by the almost simultaneous arrival of Murray's troops from Quebec and other contingents from New York. Murray's force had been so attenuated by the Quebec campaign that composite battalions had been formed; the remains of the 48th were joined with those of the 15th, the 58th with the 28th. The whole force was divided into two brigades, one commanded by Burton, the other by Howe. After the campaign Murray wrote of Burton, 'his activity and zeal were conspicuous'; and of Howe it was said, 'when the war was over, no officer had a more brilliant record'.

CHAPTER 3

The West Indies and Gibraltar

SIR John Fortescue wrote of service in the West Indies as being 'held in horror and loathing'. Unfortunately, during the rest of the Seven Years War, the American Revolution and the French Revolutionary Wars, successive British governments sought to strike at French power through numerous, largely fruitless, campaigns in those islands; a policy that signed the death warrants of many British infantrymen.

The 58th only served there twice; the first occasion resulted in the greater part of the Regiment being captured at sea; the second, though it brought them a battle-honour for the capture of Martinique in 1794, ended in their complete decimation after only a year's service. The 48th, on the other hand, had the misfortune to spend a total of fourteen years in that terrible theatre.

Havannah, Martinique, Domenica, Grenada, St Lucia, – those musical names can have struck no answering chord in the mind of the soldier. To him they signified long, sweltering days of tedium in undermanned garrisons, the constant threat of French attack or negro insurrection, and worst and most pestilential of all, the dreaded yellow fever or 'black vomit'. At the siege of Fort Moro in 1762, at which the 48th assisted, British losses from enemy action were some 1,000 killed and wounded; over 5,000 died from sickness. In 1779 the island of Grenada was supposedly garrisoned by five companies of the 48th; when 5,000 French troops attacked, only ninety men were fit enough to fight. In 1796 the 48th left England 847 strong for the capture of St Lucia; eighteen months later only fifty men remained; battle casualties had only accounted for the loss of thirty. Four times between 1760 and 1800 the 48th sailed to the West Indies; four times a pitiable remnant returned, their ranks decimated by disease. Their experience was typical of many regiments. Seldom can there have been a time when Government has been so prodigal with soldiers' lives.

It was not as though soldiers were in plentiful supply. In England and Ireland, to which the unhappy regiments returned from time to time, life was so wretched for the soldier at this period that not only were recruits hard to find, they were almost impossible to keep. The soldier's pay was so inadequate and the stoppages to which he was subject so heavy that, in Fortescue's words, 'the only alternatives open to the private soldier were to desert or starve'. It is not surprising that, under these conditions, when the 48th were inspected in Dublin in 1772 before returning to the West Indies, Lord Townshend, Lord Lieutenant of Ireland, reported, 'a very indifferent Regt, ye men are bad-sized and slovenly'. In the same report, Townshend's own Regiment, the 28th, received high praise; but then so had the 48th the previous year when inspected by their own Colonel, General William Browne, who declared the men to be of 'good size, well made and appeared very clean under arms'!

A pittance for their services was not the only abuse to which soldiers of this period were subjected. Even drummers, who rated higher in the military hierarchy than the ordinary private, had some curious duties, witness the standing orders for the Drum-Major of the 58th in 1792: 'He is to see the Officers' necessary house cleaned as often as there may be occasion, by the Drummers, it being their duty from the custom of the Army.'

Officers, too, were often unsatisfactory. Pte Aytoun of the 58th records that the company commanders 'were continually on leave of absence' and when the Regiment drilled, 'the soldiers led the officers who depended implicitly on the right- or left-hand man of the company for direction.' In 1771 Capt Osborne of the 58th fought a duel with his commanding officer, Lt-Col Burgoyne, on account of a notice accusing the latter of being lacking in personal courage. Burgoyne, in fact, appears to have been a competent officer, for his superior at Gibraltar wrote that the 58th was 'the best disciplined and best appointed regiment I ever saw come from Ireland'.

Burgoyne had succeeded Howe in 1764. By 1776 the latter was proving one of the more successful British commanders in the American War of Independence. But even Howe's fine record and his successes in America did not spare him from the cynicism and denigration that beset the Army in that epoch: 'Howe shut his eyes, fought his battles, drank his bottle, had his little whore and fought again'. In 1778 Howe, who in any case doubted the justice of the war, resigned the chief command in America.

Four years before he had been involved in one of the few innovations in a generally stagnant period. In 1774 Light Infantry companies had been officially authorized for every regiment of foot, and Howe, doubtless as a result of his experience in Canada, had been entrusted with their training. Another innovation was the proposal, in 1782, that each regiment should be affiliated to a county, and the Colonels of regiments were asked for their preferences. The replies in general showed little enthusiasm for the scheme; some rejected it with contempt. The 48th's reply is not on record but the 58th elected Warwickshire (seemingly the most popular as four other regiments also chose it), Lancashire, or else the title of 'The Black Regiment', deriving presumably from their facings. In the event the 48th were linked to Northamptonshire and the 58th, by an odd coincidence, to the neighbouring county of Rutland.

It would seem doubtful whether the officers and men of the 58th had much choice or interest in this matter, for at the time they were engaged in one of the few military operations of the period that shed any lustre on British arms – the Great Siege of Gibraltar by the combined forces of Spain and France. When the siege began in 1779, the 58th had already been in the garrison for nine years. With four other British regiments,* a Hanoverian brigade, and men of the Royal Artillery and Engineers, they were to endure nearly four more years, of blockade and bombardment, disease and destruction.

To fully appreciate the magnitude of this achievement, it is important to understand the topography of Gibraltar. Not quite three miles in length from north to south, and only three-quarters of a —

---

* The 12th, 39th, 56th Foot, and the 72nd Manchester Volunteers. The garrison was reinforced by the 73rd in January 1780, the 97th in March 1782 and the 25th and 59th in October 1782. In 1779 the garrison had 5,382 soldiers and 760 sailors; by 1782 it had risen to about 7,000.

mile broad at its widest point, it is joined to Spain by a flat, sandy isthmus, a mile long and half a mile wide. The Rock itself, rising 1,400 feet above the sea, falls precipitously into the Mediterranean on the eastern side but on the west there is a long, narrow shelf between the lower slopes and the shore. Along this shelf lay the town, barracks and harbour. From the north face of the Rock, along the western shore to Europa Point in the south, stretched the batteries, bastions, walls and palisades of the defences. Inside this cramped arena, the garrison and inhabitants lived and fought for 1,322 days.

For the first twenty-two months the Spaniards endeavoured to starve the fortress into submission by blockade, while simultaneously constructing their siege works across the isthmus. Rationing began early; one of the first economies made by the Governor, General George Eliott, was that soldiers should no longer powder their hair, in order to conserve the flour used for that purpose. The wits claimed that this edict was inspired by Eliott's liking for puddings. As early as November, 1779, civilians were fighting each other for food. Sgt Samuel Ancell, Clerk of the 58th, whose journal provides a valuable record of the siege, spoke ruefully of the soldiers' food: 'Our provision is chiefly salt beef and pork. Vegetables are scarce and dear; and what are sold are no better than the rubbish of a dunghill in Britain.' Although three major convoys managed to reach the Rock during the siege, the shortage of food caused incessant hardship and scurvy became rife. By the winter of 1780 hunger and disease posed a greater threat than the Spaniards.

The British batteries had opened fire in September, 1779, and kept up a regular cannonade against the Spaniards. However, it was not until April, 1781, that the retaliation came. As the inhabitants hastened joyfully to greet a convoy that had just sailed in, a thunderstorm of shot and shell, fired from 114 heavy guns, burst over their unsuspecting heads. As the drums beat to arms, Sgt Ancell abandoned the letter he was writing and hurried to his alarm post; later he returned to describe his experiences:

'One minute a shot batters a house about your ears, and the next a shell drops at your feet; here you lay prostrate, waiting the mercy of the explosion. If you escape unhurt, you are perfectly stunned and almost suffocated with an intolerable stench of powder. On every hand slaughtered objects lay before you; one loses an arm or leg, another cut through the body, a third has his head smashed and a fourth is blown to pieces. Blood, cries, groans, wounds, and contusions, in every part are to be seen and heard.'

Day after day the bombardment continued; at night enemy gunboats crept in and ensured no rest for the exhausted soldiery. Ancell continues:

'Between the land and sea fire we scarce dare close our eyes. The hurry of the times, the noise of the mortars, howitzers and cannon render the mind so confused; shot and shells are my near companions; smoke and wounded brother-soldiers are constantly in view; we have heavy duty, hard watchings, and little rest.'

And so it was to continue, month after month.

The Great Siege was primarily a gunner's and sapper's war. The Royal Artillery trained 180 infantrymen as additional gunners, and sixty-six of the best marksmen of all the regiments were formed into a company of snipers. But for most infantrymen it was four years of guards, picquets, working parties to build and repair the fortifications, carrying parties for ammunition, fire-fighting and innumerable fatigues and duties. It was a hard life, requiring endurance and vigilance and with little prospect of engaging the enemy at close quarters.

Only once during the siege did infantrymen leave the defences. On the night of 26 November, 1781, General Eliott launched a sortie across the isthmus to destroy the nearmost Spanish batteries. The force consisted of the 12th Foot, Hardenberg's Hanoverians, and all the Light and Grenadier Companies of the garrison, those of the 58th and 56th forming the reserve of the force. Inside Gibraltar the battalion companies of the 58th and 39th were formed in readiness to support the sortie if necessary. In two hours the force cleared the works of the Spanish guards, spiked twenty-eight guns and mortars, destroyed and burnt the batteries which had taken fourteen months to construct, and blew up the magazines, for the loss of five killed and twenty-five wounded.

Considering the hardships the average infantryman underwent in Gibraltar, it is not surprising that discipline deteriorated from time to time. Looting, drunkenness and desertion were the chief crimes. Eliott was not by inclination a flogger or a hangman but, in the straitened circumstances of the siege, he could not afford leniency. The 58th had their share of victims of his measures; Ensign Gregory court-martialled for dicing; 'the criminal John Wild' hanged with a label inscribed 'Plunderer' on his body; James Ward executed for looting; 1,000 lashes to William Rowls for the same offence. On hearing one 58th man declare he would join the Spaniards, 'the Governor said he must be mad and ordered his head to be shaved, to be blistered, bled, and sent to the Provost on bread and water, wear a tight waistcoat, and to be prayed for in church.' But despite 'the scandalous irregularity' of the garrison, Eliott felt able to report that he had no reason 'to suspect that any care is wanting at the Post next the Enemy; the soldier is wakeful when the fumes of liquor are evaporated.'

The soldier's philosophy in good times is recorded by Ancell, who, in conversation with a man named Jack Careless, made a pious observation on their preservation from death, only to be told: ' "Our King is answerable to God for us; I fight for him. My religion consists in a firelock, open touch-hole, good flint, well-rammed charge, and seventy rounds of powder and ball. This is the military creed. Come, comrade, drink Success to British arms.' " Bad times could drive three men of the 58th to their deaths by trying to desert down a cliff face with a rope that was many yards too short.

And so the siege went on; the blockade, the artillery duel, small-pox and near-starvation, the constant watching for a Spanish assault. At last it came. Against Eliott's worn and weary 7,000, were massed 40,000 French and Spaniards. To deliver the *coup-de-grâce*, the enemy assembled ten huge floating batteries, each like 'an oblong floating hay-rick', a fleet of forty-four battleships, and land batteries of 200 heavy guns and mortars. Eighty thousand spectators collected to see the fall of Gibraltar.

On 13 September, 1782, the bombardment erupted. Throughout the long day, amidst the heat, smoke and explosions, the British served their guns from the Isthmus to Europa. Ancell watched the gunners at work, 'like Ethiopians black by rubbing their faces with their hands dirtied with powder'; other men tended the furnaces preparing the red-hot shot which alone had any effect on the floating batteries by setting them alight. Into the night the thunderous exchange continued until at last Ancell could write, 'the battle is our own'.

Depressed by the failure of the floating batteries and riven by furious controversy, the enemy commanders made no attempt to attack by land. The artillery exchange continued for several months but the Spaniards' will to take the Rock was waning. The task seemed impossible, as a ribald lampoonist observed:

> ' 'Gainst Elliot the French & the Spaniards, Combin'd
> Are throwing their Stink Pots you see from behind,
> That the Garrison's Safe you must own is no Wonder
> For all that they do is but F—t—g at Thunder.'

The siege ended on 3 February, 1783. For their services, the 58th were awarded the honour 'Gibraltar' and were subsequently authorized to bear the Arms of the town, the Castle and Key, upon their badges and appointments. Later the device became the badge of The Northamptonshire Regiment. Today, three of the Gibraltar regiments, the 12th, 56th and 58th, form part of The Royal Anglian Regiment, whose badge still bears the memorial of the Great Siege.

CHAPTER 4

The Napoleonic Wars – Mediterranean

THE 58th returned home from Gibraltar in 1784 and, except for the expedition to Martinique in 1794 mentioned briefly in the last chapter, they remained in the United Kingdom until 1798. In that year, the French expedition under Bonaparte to Egypt, with its threat to India, focused the British Government's attention on the Mediterranean. As a result the 58th found themselves again in Gibraltar, forming part of a force destined for the capture of Minorca. The island was taken with little difficulty, and after remaining in garrison there until the turn of the century, the 58th moved to Malta.

By November, 1800, Bonaparte himself had left Egypt, but a French army still remained there. To eject this force a three-pronged assault was planned; the Turks from across the Sinai desert, a British force from India to sail up the Red Sea, and a third force under General Sir Ralph Abercromby to land on the Mediterranean shore. The latter force assembled at Malta, prior to sailing for the southern coast of Turkey from where the operation was to be launched. In the forthcoming campaign, the 58th were to have the distinction of serving under Sir John Moore, then a major-general commanding a large brigade known as the Reserve.

After spending six weeks practising beach landings on the Turkish coast, the invasion fleet arrived off Aboukir Bay, the place chosen for the assault, on 1 March, 1801. Bad weather prevented an immediate attack but at two am on 8 March the troops filed down to the landing craft, fifty men to a boat. By three-thirty the boats were filled and were rowed quietly towards the off-shore rendezvous where they were formed in line. The initial assault had been entrusted to Moore's Reserve, a Guards brigade and Coote's brigade. The Reserve was divided; on the right, under Moore's personal command, were the

1. A Private in the first uniform of the 48th, 1742

2. A Grenadier of the 48th in marching order, 1751.

3. A Grenadier of the 58th, from the set of water-colours illustrating the Clothing Warrant of 1768.

4. Lieutenant-Colonel George Duckworth who commanded the 2/48th at the crossing of the Douro and the 1st/48th at Albuera, where he was killed.

5. *The Sortie from Gibraltar, 27 November, 1781. The Light and Grenadier Companies of the 58th are in the foreground.*

6. *The landing at Aboukir Bay, 8 March, 1801.*

7. *The Battle of Talavera, 27–28 July, 1809.*

8. *The storming of Badajoz, 5 April, 1812. Captain Tonyn of the 48th is said to have been one of the first to ascend the ramparts.*

9. *Privates of the Grenadier and a Battalion Company of the 58th, about 1840.*

40th flank companies, the 23rd and 28th; on their left came the 42nd, 58th and Corsican Rangers * under Brig-Gen Oakes. To the left of Oakes' men were the Guards.

At eight am the boats pulled for the shore, covered by two gun-boats. On a headland at the northern end of the bay lay the castle of Aboukir with guns enfilading the beach. In the centre, opposite Moore's wing, was a dominating sandhill and to the south of it a mass of dunes covered in scrub, behind which the French lay concealed.

As the range closed, the French guns opened fire, churning up the water and drenching the men in the boats. When the landing craft approached the shore, the French gunners changed to grape-shot while their infantry advanced to the water line and delivered volleys into the crowded boats. Despite the inferno, the British rushed ashore and formed line by battalions. Moore immediately led his wing straight up the central sandhill and drove the enemy from the top. On the left the 42nd Highlanders were charged by cavalry on landing, but the 58th formed up rapidly and repulsed the horsemen with volleys. The two regiments then advanced to clear their front. Beyond the 58th were the Coldstream and 3rd Guards but, as they were forming, again the French cavalry hurled themselves down from the dunes. Sir Robert Wilson, an officer of the expedition, described what happened: 'This unexpected attack caused a momentary disorder, but the 58th regiment, formed already on their right, by their fire checked the enemy, and gave time for the Guards to present a front, when the cavalry again retreated with considerable loss.' There followed some skirmishing among the dunes but within two hours the beach-head was secure and the rest of the force was able to land. The 58th had lost Maj Thomas Ogle † and nine men killed, and forty-seven wounded but had justified their inclusion in Moore's Reserve.

The British now prepared to advance, and on 12 March moved

---

* This unit, 200 strong, had been raised in Minorca in 1799.

† One of the oldest pieces of silver owned by The Northamptonshire Regiment was a cup presented by this officer and Capt Sutton in 1799. Sutton was ADC to Abercromby in Egypt.

westwards towards Alexandria. After an action on the 13th, in which the Reserve was hardly engaged, Abercromby took up a position four miles from the city, on a ridge about a mile and a half long, with his right resting on the sea and his left on the shores of Lake Maadieh. The Reserve was placed on the right of the Army, which, as Moore realized, was the key to the position. Fifty yards from the seashore stood a gigantic, roofless building, the ruins of the Palace of the Ptolemies; at night it was an eerie place, as the 58th found when they came to occupy it. Thirty yards to their left front a redoubt was built and held by the 28th Foot, while the other regiments of the Reserve were placed in rear of these two positions. To the left across a small valley were the Guards and the remainder of the force continued the line to the lake.

An hour before dawn on 21 March the troops stood to arms, as was the daily custom. Suddenly, on the far left, a fusillade of shots rang out. The picquets of the Reserve peered more intently into the silent darkness in front of them. A few minutes passed; then they heard the dull tramp of feet advancing through the sand. Firing their muskets they retired on the main position. Now that surprise was lost, the French broke into cheers and, with drums beating, advanced rapidly on the redoubt and the ruins. Sir Robert Wilson described this attack:

'One column directed itself upon the ruins where the 58th were posted, the front of which was considerably more extensive than the front of the regiment; but some parts of the wall still standing, it admitted of the regiment's dividing itself, but scarcely notwithstanding did the troops fill up the different openings. Colonel Houston, who commanded, faintly perceived the column of the enemy advancing with beat of drums and huzzas; but fearing lest the English picquets might be preceeding, he allowed it to approach so close that the glazed hats were clearly distinguished, when he ordered the grenadiers to fire, which was followed by the whole regiment and repeated with several rounds.'

The 58th's well-directed volleys forced the French column to retire some distance, but then, joining with another, they again advanced through the smoke and darkness to try and surround the redoubt. Wilson's narrative continues:

'The 28th Regt stationed there opened a heavy fire on that part of the enemy which attempted to storm the redoubt in front; but the main body of the two columns now joined to a third, forced in behind the redoubt, and whilst some remained to attack it in rear, the rest penetrated into the ruins. Colonel Crowdjye,* who commanded the left of the 58th, observing their advance through the openings, wheeled back two companies, and after two or three rounds of fire advanced on the enemy with the bayonet. At this moment the 23rd regiment appeared to support and the 42nd also advancing to cover the opening on the left of the redoubt, so cut off the troops which had entered, that after a severe loss they were obliged to surrender. The 28th regiment had presented, as well as the 58th, the extraordinary spectacle of troops fighting at the same time to the front, flanks and rear.' †

The dawn attack against the vital sector of the British line had been held, and although, as the daylight increased, the fighting continued, every French assault was repelled along the line. By ten am the French had withdrawn. Though heavily engaged, the 58th's casualties were slight, losing only three killed and twenty-one wounded, due mainly to the cover provided by the ruins. An historian of the campaign, General Sir Henry Bunbury wrote: 'The cool intrepidity of the 28th and 58th Regiments baffled the furious and protracted assault of the confident Republicans.'

Abercromby having been killed, the command now devolved upon General Hutchinson who, deeming his force not strong enough to besiege Alexandria, decided to isolate the town and clear the rest of the country first. Accordingly a small force of the 40th flank companies, the 58th and a detachment of mounted riflemen was despatched across twenty-five miles of desert to capture Rosetta, near the mouth of the Nile. This having been accomplished, Hutchinson moved up with reinforcements and advanced along the Nile towards Cairo, reducing the French garrisons en route. Despite temperatures of 120°F in the shade, Cairo was successfully invested and on 9 July

---

* Wilson means Crougey, the second Lieutenant-Colonel of the 58th.
† For this incident, the 28th, later The Gloucestershire Regiment, was awarded the distinction of wearing an additional badge on the back of their headdress. It seems strange, from Wilson's account, that the 58th were not granted the same privilege.

the French surrendered. It now remained to return and besiege Alexandria, which eventually capitulated on 2 September. The 58th had made a successful start to the new century, taking part in every stage of the campaign, and were henceforth entitled to bear the Sphinx and the word 'Egypt' upon their Colours and appointments.

The 58th returned to the United Kingdom in 1802, as did the 48th who, since 1798, had also been in the Mediterranean in Gibraltar, Minorca, and at the capture of Malta. The peace which followed the Treaty of Amiens in 1802 was soon broken by Napoleon who, the following year, began preparations for the invasion of England. To counter this threat, the British Army was increased. As a result both the 48th and 58th received second battalions; the 2/48th forming at Manchester and the 2/58th in Ireland. Both 48th battalions were to spend the next three years in the British Isles, while the 2/58th were destined for six years in Jersey.

In 1805 the 1/58th returned to the Mediterranean and in the next year found themselves as part of the garrison of Sicily, a vital base for the Mediterranean fleet. Rather than remain passively on the defensive, Maj-Gen Stuart, the British commander, determined to attack the French under General Regnier in southern Italy. Stuart adopted the eighteenth-century practice of massing the flank companies into separate battalions; thus, when his four brigades landed in Calabria on 30 June, the 1/58th Light Infantry formed part of an eight-company Light Battalion in Kempt's brigade with the Corsican Rangers; the Grenadier Company was in a similar battalion brigaded with the 27th Foot under Cole; and the balance of the 1/58th, with Watteville's Regiment,* formed Oswald's reserve brigade.

On 4 July Stuart's 5,000 advanced against Regnier's 6,000 who were defending a strong position near the village of Maida. When he saw Stuart had only eight battalions, Regnier decided to attack first. The opposing forces were both echeloned back from the right and left respectively so that Kempt's light infantry came into action

---

* This was a foreign regiment, mainly Swiss, in the British service. Formed in 1801, it had served in Egypt and was one of the best of the foreign regiments. Disbanded 1816.

first, meeting French light infantry of the 1st Légère, a 'superb brigade', to quote General Bunbury. The clash was later described by Stuart in his despatch:

'The two Corps, at the distance of about one hundred yards, fired reciprocally a few rounds, when, as if by mutual agreement, the firing was suspended, and both Armies marching forwards by order to close combat, in awful silence they advanced towards each other until their bayonets began to cross; at this momentous crisis the enemy became appalled, they broke and endeavoured to fly, and they were overtaken with the most dreadful slaughter.'

The defeat of the 1st Légère halted the French advance and although Cole's Grenadiers had a brisk battle on the left, Oswald's brigade came up to threaten the flanks, and the French were driven from the field. Two days later Stuart advanced in pursuit southwards to the toe of Italy, and on 21 July the French garrison in Calabria surrendered. The troops then returned to Sicily.*

The 1/58th remained in Sicily for the next six years, on constant watch against invasion of the island. The monotony of garrison duty was relieved by an abundance of cheap Sicilian wine but over-indulgence resulted in a lowering of discipline and morale; this, however, was alleviated by occasional forays against the Italian coastline.

One such occurred in September 1808 when Capt O'Brien's company of the 1/58th, with 250 men of the Malta Regiment, attacked the harbour of Diamante in Calabria. While the Navy bombarded the town, the infantry landed at dawn half a mile to the north. Advancing through the undergrowth, the troops pushed back 400 Civic Guards and some French infantry, entered the town and captured a quantity of enemy shipping.

While the 1/58th performed their irksome task in Sicily, the Peninsular War was raging in Spain. In 1812 it was decided to send a British force to the eastern province of Catalonia to assist the Spaniards, whose operations tied down a large number of French troops who would otherwise be available for use against Wellington

* They embarked at Reggio, very near to where the 58th landed for the invasion of Italy in 1943.

in the west of the country. Regrettably the expedition, in which the 1/58th were included, suffered from the inertia of its commander, General Murray. A few inconclusive actions were fought, culminating in the investment of Barcelona in April, 1814, when the fighting ended.

Meanwhile, across the Atlantic, the United States had invaded Canada. The end of the war in Spain freed troops to deal with this new threat, and thus, after fifty years, the 58th returned to the scene of its earliest battles. They were engaged at the Battle of Plattsburg in September, 1814 but their stay in Canada was to be short-lived. In the early summer of 1815 news of Napoleon's escape from Elba reached Canada and the 1/58th were among the reinforcements sent to Europe. Arriving too late for Waterloo, they marched to join the Army of Occupation in Paris, where they remained until the end of the year, before returning home.

CHAPTER 5

The Napoleonic Wars –
The Peninsula

IT is now time to return to the parts played by both battalions of the 48th and the 2/58th in the Peninsular War. The 48th fought from 1809–1814 and the 2/58th in the battles of the last two years. To cover fully the services of three battalions during five years' hard campaigning is manifestly impossible in a history of this scope; it will be best, therefore, to select, out of the eighteen battles, sieges and actions they fought in, those in which a battalion of the Regiment achieved singular distinction.

The 2/48th, under Lt-Col Duckworth, was first in the field. At the Battle of Oporto in May, 1809, as part of Hill's brigade, they seized and held a bridgehead across the River Douro, thus drawing Marshal Soult's force away from the best crossing place, allowing the main body of Wellesley's army to cross unmolested. Having defeated Soult, Wellesley next turned his attention to Marshal Victor's army, which had meanwhile withdrawn from Portugal to Talavera.

The 1/48th joined the field army on 20 July from the garrison at Gibraltar, having seen little, if any, fighting in the previous twelve years. Their performance in the ensuing battles is therefore a tribute to the high standard of discipline and training achieved by the Commanding Officer, Lt-Col Charles Donellan. 'Old Charley', as he was familiarly known, was 'an excellent, though somewhat whimsical officer'; his whimsicality manifesting itself in a personal devotion to the cocked hat, powdered hair, white buckskin breeches and jack boots of the previous century. Eccentric in appearance he may have been, he nevertheless maintained discipline without recourse to flogging and was much loved by his men.

In the position adopted by Wellesley at Talavera, the two 48th battalions, although in different brigades,* were both in Hill's Second Division, placed on the left of the British line to defend the Medellin hill; to the south were posted the First, Third and Fourth Divisions along the west bank of the Portina brook, with the Spanish army extending the three-mile long line to the town itself. Holding the Medellin was vital for, if captured, its height would dominate the rest of the British position. It therefore became vulnerable when Hill's Division, on the evening of 27 July, took ground, not on the crestline as intended, but half a mile to the west.

At about ten, as the men were settling down for the night, firing suddenly broke out around the crest of the Medellin. Under cover of darkness three French columns had advanced up the hill and were now on the top. Hill immediately ordered Stewart's brigade to clear the crest. Ensign Edward Close, a young officer of the 1/48th, described his first sight of the enemy: 'We had scarcely formed line when a blaze of fire from a body of grenadiers gave us the first information of the situation of the enemy. It was so dark we could hardly see, when the musketry proclaimed the firers.' The 29th and 1/48th surged forward through the darkness, fired a volley at point blank range and hurled the enemy off the hill.

The Division was now reformed in its correct position; with Stewart's brigade on the left and Tilson's on the right, the two 48th battalions lay side by side in the centre of Hill's line. Few men got much sleep that night; the clamour of French preparations for the coming battle agitated the picquets and the men frequently had to stand to arms. When dawn broke, the growing light revealed rank upon rank of French troops, guns in front, infantry columns behind and a mass of cavalry in the rear, with further reserves approaching from the distance. Practically the whole of Victor's 40,000 men were ranged against Wellesley's 20,000, with only a handful facing the Spaniards on the right.

At five am a signal gun fired to herald a thunderous cannonade

---

* The 1/48th was in Stewart's brigade with the 29th and the Battalion of Detachments (a complete unit of men from battalions left in Portugal by Moore in 1808); the 2/48th was in Tilson's brigade with the 1/3rd and 2/66th.

10. *The 58th and sailors from H.M.S.* North Star *and* Hazard *attacking the Maoris at Mawe, New Zealand. The 58th are wearing forage caps and shell jackets. A water colour by Major Cyprian Bridge, 58th.*

11. *Officers of the 48th in the Crimea. The Colonel, B. Riky, is wearing the old coatee, others are in the new double-breasted tunics introduced in 1855, while the rest are in shell jackets.*

12. *Men of the 48th on the ranges in Malta in 1857.*

13. *Group of officers of the 58th at Sunderland Barracks, 1861. Seated in the centre is Major R. C. Whitehead, who commanded the 58th in the Zulu War and was Colonel of The Northamptonshire Regiment from 1897 to 1910.*

from the whole French artillery of fifty-four guns. As the shot rained down on the Medellin, Wellesley ordered Hill's Division to retire behind the crest and lie down. Soon the hilltop was shrouded in smoke and dust. Eyes and ears were strained to penetrate the murk and noise. Suddenly a rattle of musketry revealed the skirmishers of the light companies as they turned to fire before continuing their orderly retirement to the main position. Behind them, through the smoke, tramped nine French battalions, formed in three dense columns each about sixty files across and twenty-four ranks deep; the southernmost column was advancing directly on the centre of the two 48th battalions; the middle at a point where the left of the 1/48th joined the Battalion of Detachments; while the northern column seemed to be moving round the flank of the 29th.

When the French were a hundred yards from the crest, Hill ordered his brigades to stand up and advance. The six battalions marched forward in a line only two ranks deep to within ninety yards of the French, halted, and opened volley-fire by platoons. The ensuing three minutes demonstrated the superiority of line over column. The line formation permitted every man to use his musket, whereas only the first four ranks of the column could bring fire to bear. Reeling under the volleys, the French columns struggled to deploy into line, but the fire was so crushing that first the centre column, then the southern, began to crumble and make for the rear. Seeing this, Stewart ordered a charge. Leith Hay of the 29th wrote:

> 'With one tremendous shout the right wing of the 29th, and entire battalion of the 48th, rushed like a torrent down, bayoneting and sweeping back the enemy to the brink of an insignificant muddy stream. In the pursuit all order was speedily lost. The men advanced in small parties, destroying those of the enemy who had not ensured their safety by flight.'

Having routed the first French attack, the victorious men of the 29th and 1/48th returned to the Medellin. It was not yet seven o'clock.

A lull ensued for the rest of the morning until one pm when the French came on again. This time the attack fell on Campbell's Fourth Division on the extreme right, but his men held firm. As they did so, the main French onslaught, 15,000 men in two waves, each of twelve battalion columns, was launched in the centre against Sherbrooke's 6,000 men of the First Division; this had the Guards' brigade on the right, Cameron's brigade next, and two King's German Legion brigades on the left. The French deployment sent the twelve battalions of the first wave, each with a frontage of only thirty men, straight into the rolling volleys of Sherbrooke's eight battalions ranged in a single two-deep line. The result was inevitable. The heads of the columns were blasted away and Sherbrooke's men charged forward with the bayonet in headlong pursuit of the broken enemy. Cameron halted his brigade just beyond the Portina brook and reformed, but the Guards and K.G.L., all order lost, many with unloaded muskets, surged on straight into the musketry of the French second wave, behind which the first wave was rapidly reforming.

In minutes victory had turned into crisis. As the Guards and Germans reeled back in disorder, they carried Cameron's men with them and the two French divisions advanced on the gaping hole in the English line. Disaster threatened, but high on the Medellin Wellesley had foreseen the danger. Mackenzie's brigade,* the only reserve, was moving forward towards the right of the gap but they would not be enough. The Medellin was again under threat of attack and Wellesley could spare only one battalion from it. The 1/48th was the strongest on the hill and he ordered them down into the breach.

Donellan's voice, 'as he gave the word of command along the line of his battalion was like a match to the gun'. Doubling down the hill, the battalion wheeled back by companies to allow the retreating Germans through and then, in Napier's words, 'resuming its proud and beautiful line', went forward like a wall to block the triumphant advance of the French right. At this moment the Colonel was mortally wounded and, raising his old tricorne, he courteously invited Maj Middlemore to lead the 48th to the charge. At a range of fifty yards their volleys crashed out, one every fifteen seconds from each

---

* Of the Third Division. The other brigade of this Division, Donkin's, was positioned on the Medellin; it had suffered severe casualties on the morning of the 27th.

of the twenty half-companies. The French columns withered and, with Mackenzie's brigade firing furiously on the right and the First Division now reformed and coming up in support, the crisis was over.

Wellesley, soon to be Viscount Wellington, wrote in his despatch: 'The battle was certainly saved by the advance, position and steady conduct of the 48th Regiment'. Pte William Coles of the 48th wrote in a letter: 'We was almost sweltered with heat and fatigue. I shall never forget it, as I believe it was such a day as has not been for bloodshed since the memory of man.'

Pte Coles' surmise was not extravagant. Nearly 12,000 French and British soldiers were killed and wounded. His own Regiment had lost 176, including the much-loved Colonel of the 1st Battalion. Yet in just under two years time the 48th was to see one battalion practically annihilated and the other suffer crippling losses.

On 15 May, 1811, a force under Sir William Beresford arrived at the little town of Albuera to halt Soult's corps advancing to relieve Badajoz. Late in the evening Beresford was joined by a Spanish force and the Fourth Division was hurrying through the night to reinforce him. He took up a position to the west of Albuera with a Portuguese division on the left, the Second Division in the centre opposite the town, which was occupied by a King's German Legion brigade, and the Spaniards on the right. The 2/48th was in the same brigade as at Talavera but had been joined by a fourth battalion, the 2/31st, and a new brigade commander, Colborne. In the 1/48th's brigade, now under Hoghton, the 1/57th had replaced the Battalion of Detachments. Lt-Col Duckworth had left the 2nd Battalion to assume command of the 1st, handing over the former to Maj Brooke.

Early on the 16th the French feinted against the town but shortly afterwards, nineteen battalions began a massive attack from the south to roll up Beresford's position from the flank. He ordered the Spaniards to wheel to meet this onslaught and sent the Second Division to their aid. However, the Spanish commander only deployed four battalions which, though fighting more courageously than usual, were beginning to yield under the pressure when Colborne's brigade came up on their right flank.

Ensign Close, now in the Light Company of the 2/48th, takes up the tale:

'The morning was heavy and misty. We got orders to move off to the right, which we did in column of companies and arrived just as the French were driving in the Spanish sharpshooters. A very heavy cannonading was scattering destruction amongst us at this time. Our brigade began to form line, the Buffs on the right, 2nd Battalion 48th next, and the 66th Grenadiers being all that could be said to be formed and the 66th moving in echelon to form. Before the 31st Regiment could open out, we were bayonet to bayonet with the enemy. We found the enemy, all grenadiers, pouring in a dreadful fire up the hill. Two or three volleys were fired by our regiment, when, irregularly formed as we were, we charged. The left column of the French became opposed to the left wing of the Buffs and our right. Their centre column faced our two left companies and the 66th Grenadiers. Their right found its way to our rear. Although we were literally cut to pieces, we stood on the hill like extended light infantry. The French left column was broken and was the only part which stood the charge. They remained as if powerless until bayonetted by our men. The centre column gave way as soon as we charged. We kept advancing until we received an order to retire, when facing about we beheld a mass in our rear firing away handsomely. Those of the other columns resumed the fight and commenced a murderous fire.'

Suddenly a thunderstorm burst over the battlefield. The rain and hail fell in torrents, blotting out vision and rendering the muskets useless. Under cover of the storm and the battle smoke, two French cavalry regiments, one of them Polish lancers, charged the right flank of the Buffs as they stood in line engaging the enemy infantry. The line formation against infantry was superb; against cavalry it was fatal. Rampaging down the Buffs' line, through the 2/48th and the 66th, the cavalry galloped, spearing and hacking the living as well as the wounded lying in the mud. Only the 2/31st, at the north of the line, had time to throw themselves into square and resist the savage Poles.

Ensign Close was one of the few to escape:

'They rode through us in every direction, cutting down the few that remained on their legs. There was nothing left for it but to run. In my flight I was knocked down by some fugitive like myself. Whilst on the ground I was ridden over by a number of Lancers, one of whom passing close to me was about to save me the trouble of recording this event, when a Spanish dragoon rode up to him and struck him with his sabre, which brought him over his horse's head. I then got up and ran again, when I found myself between the French and the 4th English Dragoons, who were in the act of charging. On arriving at the foot of the hill, I found the Fusilier Brigade.'

Though the Poles lost a quarter of their strength in the charge, Colborne's brigade had been annihilated and, save for the 31st, ceased to exist as a fighting force. But now Hoghton's and Abercrombie's brigades came up to relieve the Spaniards who, though still fighting, had suffered severely. The six battalions, joined by the 2/31st, formed line across the ridge to engage the 8,000 French infantry; the great bulk of whom, supported by artillery firing grape and canister, and formed in a solid mass, was opposed to Hoghton's three battalions: 57th in the centre, 29th and 1/48th to right and left. The rain had stopped and now began the great fire-fight of Albuera.

An officer of the brigade, Leslie of the 29th, described this murderous conflict:

'A most overwhelming fire of artillery and small arms was opened upon us, which was vigorously returned. There we unflinchingly stood, and there we fell, our ranks were at some places swept away by sections. Our line at length became so reduced that it resembled a chain of skirmishers; while from the necessity of closing in towards the colours, and our numbers fast diminishing, our right flank became still further exposed.'

For an hour the dreadful carnage went on, in an area no larger than a cricket field. Hoghton was killed, the three commanding officers were all dead or severely wounded, four-fifths of the brigade were down, but still the volleys slammed out as the men edged in to the centre and imperceptibly advanced the line closer and closer to the French. At last, when the remnants were only twenty yards from the enemy, rescue came out of the smoke on the right flank.

Cole's Fourth Division, having arrived earlier in the morning, had been held in reserve but now his Fusilier brigade marched up the hill and launched themselves in a magnificent attack against the French left. While the exhausted and blackened remnants of Hoghton's regiments leant upon their muskets in the rain amongst their rows of dead, the Fusiliers drove the French headlong from the field.

Out of 1,650 officers and men, Hoghton's brigade had lost 1,044 but by their courage and discipline had held a far superior force in check until relief could arrive. Moyle Sherer of the 34th, who witnessed this great struggle from the ranks of Abercrombie's brigade, paid the survivors this tribute:

> 'A captain commanded the brigade; the 57th and 48th regiments were commanded by lieutenants; and the junior captain of the 29th regiment was the senior effective officer of his corps. Not one of these regiments lost a man by the sabre or lance; they were never driven, never thrown into confusion; they fought in line, sustaining and replying to a heavy fire, and often charging; and when the enemy at length fled, the standards of these heroic battalions flew in proud, though mournful triumph, in the centre of their weakened but victorious lines. I know of little which can compare with, nothing which has surpassed, the enthusiastic and unyielding bravery, displayed by these corps on the field of Albuera.' *

The captain who brought the brigade out of action was Gilbert Cimitière of the 48th; a French emigré, he had been commissioned from the ranks of the 14th Foot in 1796. The pre-battle regimental strength of the two 48th battalions had been sixty-two officers and 887 men; together they lost thirty-nine officers and 584 men. With such crippling casualties, the Regiment could no longer maintain two battalions. The privates of the 2nd Battalion were transferred to the 1st, whilst the cadre of the former returned home; it was disbanded in 1814.

The 1st Battalion fought on through the war; at the sieges of Ciudad Rodrigo and Badajoz and the great victories of Salamanca and Vittoria. By July 1813, now with the 3/27th in Anson's brigade of the Fourth Division, they were fighting amongst the mountains

---

* An exhortation by their Colonel gave the 57th, later 1st Middlesex, the nickname of 'Die-hards'; a sobriquet equally applicable to the 29th and 1/48th.

of the Pyrenees. On the fourth anniversary of Talavera, at Sorauren, Wellington again used the 48th at a critical moment of the battle. As four French brigades were forcing back two allied formations, he threw Anson's brigade against their left flank. In Napier's words:

> 'Wellington sent the 27th and 48th from the higher ground in the centre against the crowded masses, rolling them backwards in disorder, and throwing them one after the other violently down the mountain side; and with no child's play, for the two British regiments fell upon the enemy three separate times with the bayonet, and lost more than half their own numbers.'

Five days later the 2/58th also distinguished themselves in a charge against superior numbers. Joining the field army in 1812, they had served at Salamanca, the siege of Burgos, and Vittoria; now, as a result of their low strength, they and the 2/24th formed the 3rd Provisional Battalion in Barnes' brigade of the Seventh Division. With the 6th Foot they charged unsupported through the fog at Echelar against two French divisions and drove them from the heights. Wellington, with unusual warmth, wrote that the charge was made with 'a regularity and gallantry' which he had seldom seen equalled. Fortescue's verdict on the various battles of the Pyrenees was that 'the Fourth Division and Barnes' brigade were the heroes of the fights'.

Thus in the final battles to clear the French out of Spain, the 48th and 2/58th had played prominent parts. There now remained the battles in southern France, Nivelle, Nive, Orthez and Toulouse, in all of which both battalions continued their good work. When peace was signed in April 1814, the 48th sailed to Ireland and the 2/58th to England, where they remained throughout the Waterloo campaign. In December 1815 the 2/58th was disbanded.

Eleven battle honours on their Colours, granted to the 48th and 58th for their services in the war, testify to The Northamptonshire Regiment's claim to be included among what Sir Arthur Bryant has called the Peninsula's 'great, undemonstrative regiments of the Line'. But the most prized honour was always Talavera. The anniversary of the battle was celebrated annually by the Regiment, items of Regimental silver commemorated it, and its name bore a prominent place on buttons and badges. Finally, to honour the memory of the officers and men who had done so much to gain the victory, a solemn toast would be drunk each Talavera Day: 'To the 48th'.

CHAPTER 6

Savage Warfare

Anew task in a new country awaited the 48th after the Napoleonic Wars. In 1817 they sailed to Australia, where they spent seven years maintaining order in the convict settlements and hunting down the bushrangers who harassed the colonists. This was followed by the Regiment's first tour of duty in India; nine peaceful years until 1834 when the 48th joined an expedition to subdue the tyrannical Rajah of Coorg.

This short campaign in the jungles of southern India had for the 48th echoes of the march on Fort Duquesne eighty years before, as can be seen from Lt Webber Smith's description of the fighting:

> 'We fired into bushes and trees – they fired at men; and whilst we scarcely ever got a glimpse of their dark skins, our beautiful red coats – our shakoes – our white belts and glittering breastplates* were the 'bull's eyes' of the target they could see a mile off! No men have a chance in jungle warfare with such a dress.'

Despite these difficulties, the expedition was successful and shortly afterwards the 48th returned home.

They were to see no further action until 1855 when they found themselves in the trenches before Sevastopol. Arriving in the Crimea in April, they missed the great fights of 1854 and the dreadful winter that followed; they were also spared the ruinous attacks on the Redan in June and September. Their lot was the constant watch and ward in the trenches, enduring the bombardment and repelling the Russian sorties; an inglorious task but one recognized by the award of the battle honour 'Sevastopol'.

Meanwhile the 58th had been campaigning in a land that had scarcely seen a British soldier. On 4 April, 1845, in New South

---

* The rectangular plates with regimental devices placed on officers' sword belts, and on the men's pouch and bayonet shoulder belts at the point where they crossed on the chest.

Wales, where the Regiment had recently arrived from England, Maj Cyprian Bridge, the acting Commanding Officer, recorded in his journal:

'Received orders to hold the Head Quarters of the 58th Regt in readiness to embark on the shortest notice for New Zealand on special service, alarming accounts having just been received of another outbreak among the natives, and the total destruction of one of the principal settlements in the Bay of Islands.'

The instigator of the uprising was a Maori chief, Hone Heke, who was determined to flout British authority in the North Island. The proud and ferocious Maoris possessed considerable military aptitude, particularly for the selection and fortification of strong defensive positions known as *pahs*. Their chief weapons were tomahawks and firearms, acquiring from traders the new percussion muskets with which the British Army was gradually being re-equipped, but which as yet had not reached the troops in the Antipodes, who still had the old flintlocks. To the 58th's soldiers, such as Alec Whisker, the Maoris appeared 'frightful to look at and they have all kinds of things in there ears, some pieces of Bone and ivery, others with piece of sticks, others a great lot of worsted and i seen one with a Brass Padlock in his ear.' Cannibalism was still occasionally practised, a characteristic apparently known to the Bandmaster of the 58th, for as the Regiment landed, the band, the first in New Zealand, played '*The King of the Cannibal Islands*'.

In the fighting that followed the 58th was split up in columns and detachments and never fought as a whole regiment. A column would hack its way through the forests to a *pah* in which the Maoris were reported; endeavour to breach the immensely strong defences with the few naval guns that could be dragged through the undergrowth, and then attempt to capture the *pah* by assault before the defenders could slip away to another position. These laborious manoeuvres went on throughout 1845 until the following January, when Heke made peace after the capture of the *pah* at Ruapekapeka. However, guerrilla activity continued in the south of North Island until 1847.

Although the 58th predominated in the columns, the operations were initially commanded by Lt-Col Hulme of the 96th, but he was superceded when Col Despard and part of the 99th joined the force. Despard's abilities were gravely doubted by the 58th. Cyprian Bridge had quickly grasped the essentials of Maori warfare and, being solicitous for his men, was appalled at Despard's incompetence. Before Ruapekapeka he wrote:

'How deplorable it is to see such ignorance, indecision and obstinacy in a Commander who will consult no one or attend to any suggestion made to him and also in consequence has neither the respect nor the confidence of the troops.

Cpl John Mitchell recorded Sgt-Maj Moir's opinion of one of Despard's foolhardy schemes: ' "The chances are against us coming out of this alive, Mitchell. I look upon it as downright madness;" ' * and when yet another of the Colonel's projects failed, 'the affair was commented upon in no measured terms by the whole force'.

The endurance of the men was sorely tried by the long marches in vile weather, through thick undergrowth, and with inadequate transport. The night's bivouac afforded little rest, as Colour Sgt Robert Hattaway described:

'The camp was miserable; the ground saturated from incessant rain, the flooring of the tents a pool of mud. My three comrades and myself constructed a small tea-tree shed, with two blankets for roof protection and fern for our beds. Our accoutrements remained on us, and our arms by our side, as our proximity to the rebels necessitated every precaution against surprise. There had not been stored provisions for a winter campaign and biscuits and rum were the daily fare.'

The parade-ground gloss of the 58th, which had so reassured the colonists on landing, was soon dulled as Hattaway's description of the Grenadier and Light Companies before Ohaeawai shows:

'They were a magnificent body of men, thoroughly disciplined and trained. Their knapsacks having been placed in store, they had no change of garments, were ragged, tattered and torn, many without boots and tied on their feet with flax, their pants of many colours; blankets and greatcoats reduced in size to repair their continuations.'

Alec Whisker noted how active service made even the officers 'like

* Moir's fears proved groundless, for the attack, on Ohaeawai, was postponed owing to bad weather.

aney of ourselves', as he watched an officer trudging along with 'a gunk of mutton and Biscuit in his hand eating it as Ravenous as a cat would do a lark and a stick of Cavendish tobacco and his pipe stuck over his breastplate.'

Despite the hardships, poor leadership and defective commissariat, it is clear, from the several accounts written by men of the 58th*, that their morale and determination remained high. Hattaway records that when volunteers to lead the assault on Ohaeawai were called for, all the 58th stepped forward. Although the stockade was hardly damaged, the attack went in with dash as Cpl William Free vividly described:

'We formed up in close order, elbows touching when we hooked them; four ranks, only the regulation twenty-three inches between each rank. We waited in the little hollow before the *pa*, sheltered by the fall of the ground. We got the orders "Prepare to charge"; then "Charge". Up the rise we went at a steady double, the first two ranks at the charge with the bayonet; the third and fourth ranks with muskets and fixed bayonets at the slope. When we were within about fifty paces of the stockade we cheered and went at it with a rush, our best speed and "divil take the hindmost". The whole front of the *pa* flashed fire and in a moment we were in the one-sided fight – gun flashes from the foot of the stockade and from loopholes higher up, smoke half hiding the *pa* from us, yells and cheers, and men falling all around.'

John Mitchell of the Grenadier Company continues the story:

'We were met with such a fusillade, I can only describe it as the opening of a monster furnace. We expected scaling ladders and axes to have been brought up but alas there was but one ladder. The party was literally mown down. My nearest comrades were Capt Grant and Sergt Major Moir, we could see inside the *pa*, but could not reach the maories lying in the ditches. One of them a big fellow I could just reach with the bayonet, but could not use it with effect. He was intent on shooting me, I called Capt Grant's attention to the fix I was in, he shot the fellow with his pistol. He, Grant was killed almost immediately, Sergt Major Moir was wounded and I was wounded above the left knee. The bugles sounded the retreat, the

---

* See *Journal of the Society for Army Historical Research*, Vol. LI, page 69, 'Baptism of Fire in New Zealand, 1845' by the author.

engagement lasting about ten minutes with the loss of half the men and two officers.'

After this action, Ptes Whitethread and Pallett volunteered to return under fire to bring off wounded men, a deed which, ten years later, could have earned them the Victoria Cross.*

Great bravery by a young bugler was also shown at Boulcott's Farm, where a 58th outpost was stationed in May, 1846. When Maoris attacked in the early dawn, Drummer William Allen leaped up to sound the alarm but as he did so, a Maori hacked off the arm holding the bugle. Allen picked up the instrument with his other hand and again contrived to warn his comrades until he was cut down. By his self-sacrifice the post was alerted and the attack beaten off.

The campaign ended in July, 1847 but the 58th spent a further eleven years in New Zealand. In that time the Regiment became so closely integrated with the colony that, by the time they sailed for home, 1,000 men remained behind as colonists.†

In 1879 the 58th again sailed to the aid of a colony threatened by an equally formidable race. The annihilation, on 22 January, of the 24th Regiment at Isandhlwana by the Zulus necessitated the urgent despatch of reinforcements from England. Landing in Natal in April, the 58th joined Lord Chelmsford's force in time for the decisive battle of Ulundi.

At dawn on 4 July, Chelmsford formed his six battalions into a huge square with the cavalry and baggage in the centre. The 58th held the left rear face, being formed in four ranks, two kneeling and two standing. At about nine am the Zulus came on in a great horn formation, their flanks rapidly extending to encircle the square. The British artillery opened fire, followed, as the Zulus charged, by section volleys ‡ from the infantry. Cpl William Roe's journal describes what the 58th experienced:

* Instituted 5 February, 1856. Many of the early VCs were rewarded for saving wounded under fire.

† Amongst them were Hattaway, Mitchell and Free, who all later became officers in the New Zealand Militia. Cyprian Bridge returned as Governor after giving up command of the 58th in 1860.

‡ By this time the infantry were armed with the 0.455 Martini-Henry rifle, the first brass cartridge case rifle to be adopted by the British Army.

'The fire from the Artilary did fearful execution for we could see their Heads, legs and harms flying in the air, in all Directions. On they came till they were within range of our rifles. We opened a fearful fire upon them. Our fireing was in Vollys, and the Bullets went flying as thick as rain. They were falling down in heaps as though they had been tiped out of carts. On they still came till they were up to the Square all but about sixty, or seventy yards. You could not see many yards in front, for the Dence Clouds of Smoke from our guns. We made up our minds to fight in close quarter with our Bayonets but the enemy Began to shake in front of our fire, they halted dead for a few Seconds, then turned and flew for their lives.'

Thus in half an hour the Zulu power was broken.

After the campaign, the 58th remained in various garrisons throughout Natal and the Transvaal, happily unaware of the trials they were soon to undergo.

Although only single-shot, its Boxer cartridges (despite their tendency to jam in the ejector) permitted a high rate of fire; it was this fire, more than anything, which was to destroy the Zulus.

CHAPTER 7

Boers and Afridis

I N December 1880 the Boers in the Transvaal, then under British
rule, rebelled and laid siege to the British garrisons throughout
the colony. Although disposing of only a small force, Sir George
Colley, the Governor and GOC in Natal, determined to attempt the
relief of the besieged garrisons and accordingly advanced to the
Transvaal border. He found himself forestalled by the Boers, who
held a strong position astride the road at Laing's Nek; this could only
be attacked frontally.

At six in the morning of 28 January, 1881, his force paraded for
the attack: a mixed squadron of King's Dragoon Guards and
mounted infantry, five companies each of the 58th and 3/60th Rifles,
and some light artillery; about 1,000 men opposing 2,000 Boers
entrenched in a semi-circle of high ground which dominated the
approaches. Colley's plan was for the mounted squadron to attack
the Boer left under cover of the artillery, thus securing the right flank
of the 58th who were to make the main attack on the Boer left centre;
the 3/60th were in reserve.

Despite the enemy's numerical preponderance and the steep climb
ahead, the 58th had been ordered to take the position with the
bayonet; contemptuous of the Boers' fighting qualities, the Staff
considered that they would not stand cold steel. As the 58th advanced
in column of companies, they made a brave show in their white
helmets and red coats. For the last time a British battalion was going
into action in the old style, officers mounted, men in close order and
Colours flying.* Yet the enemy were skilled marksmen, armed with
modern rifles, entrenched and concealed.

From the start things went wrong. The artillery support, which

* Colours were never again carried in action by British infantry. The
58th Colours, presented in 1860, remained in service until 1962 and are now
in the National Army Museum.

had been making good practice on the Boer positions, stopped too soon. The mounted attack failed, exposing the right flank of the infantry. Command of the 58th's attack had been assumed by Col Deane of the Staff, who, being mounted, forced a gruelling pace up the steep slopes. As a result, by the time the 58th reached an area of dead ground, where they expected to deploy for the assault, the men were panting for breath. They were given no respite.

Capt Lovegrove, the acting second-in-command, continues the story:

> 'We moved on after a very short pause, and having got up a steep bit of ground, we came under a heavy point blank range fire. Mindful of my orders about the bayonet charge, I front formed my half battalion under fire and moved straight on. At this time I thought I saw the Boers giving way and I shouted out "They are retiring, follow me!" and the men came on. As the fire got so hot, I gave the order to charge. We were then within about forty to eighty yards of the Boers who were lying down behind stones.'

The fire was murderous from front and flank. Colonel Deane was soon killed; Maj Hingeston, the Commanding Officer, was mortally wounded as he encouraged his men; and most of the officers went down. Lt Baillie fell with the Regimental Colour, saying to Peel, his fellow ensign, 'Never mind me, save the Colours'. As Peel tried to comply he fell into an ant-bear hole and Sgt Bridgestock, thinking Peel was killed, seized both Colours and carried them out of action.

There was nothing for it but retreat. As the men went back, Sgt-Maj Murray, though badly wounded, directed fire to cover the withdrawal, being one of the last to retire. Lt Hill remained behind to try and save Baillie and later rescued two wounded men, thus earning the Regiment's first Victoria Cross.* Pte Godfrey stayed to protect the dying Maj Hingeston and he, Murray and Bridgestock were all later awarded the Distinguished Conduct Medal. Despite their terrible experience, the remains of the Regiment formed up at the

---

* He died, as Maj Alan Hill-Walker, in 1944, for many years the senior holder of the VC in the Army. One of the men he rescued, Bandsman Tuck, kept a journal of the campaign, which is now in the National Army Museum.

14. *The advance of Sir George Colley's force to the Transvaal, 1881.*

15. *Defending a redoubt during the Siege of Standerton in the Transvaal War.*

16. *Private Shanahan, 48th, 1892. He was born in the Regiment, enlisted at 14 in 1852, served in the Crimea and died in the Regiment, aged 54.*

bottom of the hill under Lt Jopp to await further orders. That the 58th had done their best is testified to by an eyewitness, Lt Marling of the 3/60th:

> 'The 58th behaved with great gallantry and retired in a most orderly way. If the command of the battalion had only been left in the hands of the Regimental officers, I firmly believe they would have carried the position.'

Meanwhile the garrisons in the Transvaal held out. At Standerton B Company, 58th, under Lt Compton, with three companies of the 94th, held the town for two months until peace was signed. Capt Saunders conducted an energetic defence with 120 men of A and C Companies at Wakkerstroom where, on 22 February, Pte James Osborne gained the second Regimental VC of this short campaign by riding under heavy fire to rescue Pte Mayes from a party of forty-two Boers. Although invested by superior forces, these and the other garrisons all resisted successfully till the end of the war; as one military historian, Maj G. Tylden, has noted, 'the defences were characterized by the very fine spirit shown by all ranks.'

The first attempt to relieve these garrisons had failed. After an engagement on the Ingogo River where the 3/60th received their demonstration of Boer skill, Colley devised a new and daring plan to turn the enemy's right by capturing the Majuba mountain which towered 1,500 feet above the Boer positions astride Laing's Nek; the mountain was picquetted by day but not by night. Colley was now reinforced by the 92nd Highlanders, who swaggered into camp on 19 February; a confident, seasoned battalion, fresh from a victorious campaign in the Second Afghan War. Unlike the 58th and 60th, whose men were young soldiers, products of the new short-service system, the 92nd were a relic of the old long-service Army.* The confrontation, between the proud, bearded Highlanders in their Indian khaki and the youthful victims of Laing's Nek and the Ingogo in red and green, can well be imagined. With a mixed force of these instruments Colley set out to seize Majuba: two companies each of the 58th and 92nd and a company of sailors for the summit, two

---

* Cardwell's Short-Service Act had been passed in 1870.

companies of the 60th and one of the 92nd to provide links between Majuba and the camp. One of the 58th was overheard telling a Highlander that he wished all the 92nd were there; he spoke truer than he knew.

The tragedy of Majuba is notorious; the scaling of the mountain by night; the high spirits of the men at dawn, with the cocky Highlanders yelling abuse at the Boers far below; the failure by Colley to entrench the position; the swift reaction by the Boers, advancing up the hill by fire and movement, a tactic then unknown in the British Army; the fierce fire fight on the summit; the sudden panic; Colley's death; and finally the sad spectacle of the survivors, 58th, Highlanders and sailors, running for their lives down the slopes they had so confidently climbed seven hours before.

Who broke first – the young soldiers of the 58th, fearful of another dose of devastating Boer musketry, or the 92nd, experienced in Asia but not in Africa, who held the parts of the summit against which the Boer attacks were made? T. F. Carter, a journalist present on the hill, wrote: 'Who it was, I defy anyone to say'. He was probably right, for by the time the rout occurred, the regiments were inextricably mixed. Herein lay the chief cause of the disaster. General Sir Ian Hamilton, then a subaltern in the 92nd, wrote: 'Why had Colley started upon an adventure so perilous with a scratch lot of soldiers and sailors instead of one homogeneous battalion?' Even the 120-strong reserve at the centre of the position was drawn from all three units. Colley had his reasons * but they resulted in a lack of cohesion, which bred confusion and later panic when the Boers reached the summit.

Inevitably recriminations followed, each regiment blaming the other. It is therefore pleasing to record the generous spirit shown by an officer of the 92nd, Lt Hector Macdonald,† who wrote to Col Bond, commanding the 58th:

* See Sir William Butler's *Life of Sir George Colley*.
† Macdonald was a famous Victorian military figure. Commissioned from the ranks for gallantry in the Afghan War, he later became a major-general in the Boer War of 1899–1902.

'The men of your Regiment under my command in the advanced koppie without exception proved themselves true and devoted soldiers; their gallantry was beyond praise, for even when they were nearly surrounded by the Boers and half their number killed or wounded and saw the others disappear from their view, they did not move or murmur. They retired by my order only and I believe Sergeant Giles to be the only survivor and I have great pleasure in naming him as deserving of recognition for his conduct.'

Majuba was the last fight of the 58th as a separate regiment. In 1872 Lord Cardwell's reforms had linked the 48th and 58th for certain administrative purposes but each had retained its own identity. The Stanley Committee of 1877 recommended a closer union of Regular regiments and the Militia on a territorial basis, and in February, 1881 a further committee* suggested that the 48th and 58th should form 'The Northampton and Rutland Regiment'. However, a shorter title was preferred and on 1 July, 1881 the 48th, 58th and two battalions of the Northamptonshire and Rutland Militia, and the Northamptonshire Rifle Volunteers became respectively the 1st, 2nd, 3rd, 4th and 1st Volunteer Battalions of The Northamptonshire Regiment.† However, in the Regular battalions old habits died hard and the Regimental numbers remained in common usage; a custom that will be observed here.

The amalgamation found the 48th engaged with Fenians in Ireland. Eleven years later they sailed for the fourth time to India and in 1897 were suddenly ordered to the North-West Frontier, where a serious uprising by the Afridis and Orakzais had broken out. In September they joined General Lockhart's force of two divisions, whose task was to advance into the Tirah Maidan, the heart of the enemy territory and an area hitherto unknown to British troops.

Fighting its way forward, the force reached the Maidan on 31 October where it encamped. The troops were initially without tents and in their thin khaki cotton they suffered, for, as Lt Gifford wrote in a letter, 'We had twenty-four hours marching in snow over our

---

* Ellice Committee on the Formation of Territorial Regiments as proposed by Col Stanley's Committee.

† The 3rd and 4th (Militia) Battalions were amalgamated in 1899. For origins of the Militia and Volunteers, see Appendix.

ankles. It freezes hard every night and we find it very cold.'

On the evening of 1 November, as they returned to camp after a day's picquetting, B Company, under Lt Bulwer, with Lts Dobbin and Knox,* went to the aid of a small detachment of the 2nd Queen's, who had been surprised by Afridis. Bulwer despatched Sgt Guy, of whom more later, to the camp a mile away to report. Guy made his way alone through the darkness and the prowling tribesmen and reached camp safely. Meanwhile B Company had attacked and after hand-to-hand fighting had routed the Afridis. Some time later, as C and K Companies returned to camp from picquetting duty, they were also attacked and surrounded. Lt Gifford wrote home: 'It looked as if all C Company would be cut up but the men at once got under the bank and replied to the shots coolly'.† The Regimental Record noted that the enemy 'kept up a desultory fire and continual howling' but was eventually driven off.

The next action of the 48th might have been the archetype for a story by Kipling, or one of the battle scenes so favoured by Victorian military painters at the height of Empire. It had all the ingredients: the wild, inhospitable terrain; small groups of khaki-clad British infantry, in the conical helmets so evocative of the late Victorian epoch, battling their way down a defile; men falling on all sides, the wounded being helped along by their comrades; and all around the fierce, savage tribesmen closing in for the kill. All this the 48th experienced at Saran Sar on 9 November, 1897.

The early part of the operation to reconnoitre the Waran valley, to the east of the Maidan, from the high pass of Saran Sar had been carried out against only slight opposition. By eleven-thirty the 48th were in position at the top of the pass, with the 1st Dorsets and 36th Sikhs holding ground to the left and right respectively. The Generals and Staffs arrived, had lunch, made their reconnaissances and by two-thirty the retirement began, five companies of the 48th under Maj Compton forming the rearguard.

---

* Later General Sir Harry Knox, KCB, DSO; Adjutant General 1935; Colonel of The Northamptonshire Regiment 1931–1943.

† Gifford was killed five days later by a sniper, while eating his dinner in camp.

The account written afterwards by Compton tells the story:

'The Companies were on a rocky cliff overhanging the pass and it was thought that as soon as the picquets on the opposite hills were withdrawn the enemy would occupy these hills and bring a fire on the road by which the Brigade was retiring. The rearguard was therefore ordered not to retire until the picquets had passed 600 yards down the road.'

When the time came, the companies retired in file at the double, A and B Companies going ahead to take up covering positions. Compton continues:

'Before the last company had got 200 yards from the ridge it had been holding, the enemy appeared on it and opened a hot fire at point blank range, resulting in several casualties. This Company (G) having no stretchers could not move until stretchers and reinforcements had been sent to it.'

Under covering fire from the 36th Sikhs on the right, this was effected and G Company rejoined the Battalion in the defile that led back to the camp. The Sikhs then continued their retirement down a parallel route.

Meanwhile the Dorsets had withdrawn towards camp and were now some way ahead of the 48th. Apparently unaware of this, Lt-Col Chaytor, commanding the 48th, ordered his five leading companies* to march straight home down the defile. This order meant, in Compton's words, that:

'Towards sunset the convoy of wounded with three very weak companies was left four miles from camp with its right flank and its rear completely exposed to the attacks of the enemy. On reaching the main *nullah* they suddenly came under a hot fire from magazine rifles at fixed sights range and many casualties occurred in all three companies.'

Gravely hampered by the wounded, which in Frontier warfare could never be left behind, the men fought desperately to get clear of the *nullah*. Sgt Guy again distinguished himself by going back under heavy fire to rescue Colour-Sgt Hull who had been wounded in the ankle, and Cpl Gray carried off the severely wounded Sgt

* The 48th had eight companies in all: A, B, C, D, F, G, I, and K.

Litchfield (a man of over six feet) from under a very accurate fire. Both men were later awarded the DCM. Gradually the wounded were brought to safety, but Lt MacIntyre, Colour-Sgt Luck and fifteen men of C Company held on too long, were surrounded and shot down to a man. Compton wrote, 'this small detachment gallantly covered the retirement of the wounded convoy and sacrificed their lives in courageous performance of their duty.'

The bravery of the men of C, F and K Companies in sticking to the wounded elicited high praise, but their plight was due to a tactical blunder and the command of the 48th came in for some criticism. On 7 December the force withdrew from the Tirah and by Christmas Day the 48th were back in Peshawar after a hard march, in bitter cold, on short rations and under constant harassment by the tribesmen.

Returning to the 58th, it may be noted that their unfortunate experience in the Transvaal War saw the birth of the Northamptons' reputation as a great shooting regiment. The paramount lesson the 58th learnt at the hands of the Boers was the need for individual marksmanship. By 1886 the Battalion had won the Inter-Regimental Rifle Shooting Cup three times in succession; by 1895 they were classified as the best shooting battalion of the seventy-two battalions at home; in 1898 all eight companies were among the prizewinners in the Evelyn Wood Cup;* and in 1899 the first four places in that competition were taken by the 58th.

Thus when the Battalion returned to South Africa in November, 1899, where war had again broken out with the Boers the month before, it was, in the words of Lt Barton, 'complete and ready for the "return match" it has longed for since Laing's Nek'. Other lessons from the earlier campaign had been learnt as Barton noted in his diary:

'Officers are to discard their swords and carry rifles or carbines to confuse the Boer marksmen. All the old Aldershot Drill Book Tactics are to be abolished and we shall adopt very extended order, men getting quickly from rock to rock, irregularity of line being sought and "regular dressing" avoided.'

* A marching and shooting competition open to 500 infantry companies.

The efficacy of the 58th's training was soon tested when, having joined the 9th Infantry Brigade of Lord Methuen's First Division for the relief of Kimberley, they came up to the Boer position at Belmont on 23 November, 1899. Lord Methuen ordered a night advance followed by a dawn attack with, on the right the 1st (Guards) Brigade, and on the left the 9th Brigade, deployed with the 1st Northumberland Fusiliers and the 58th forward and the 2nd King's Own Yorkshire Light Infantry in reserve.

Barton's diary describes the initial stages of the attack:

'The column started at two-thirty; the first shots were fired soon after four am. The bullets now began to whistle and scream while the outline of the hills was lit up with brilliant flashes. As we got near the foot of them the hail rendered it advisable to get flat on the stomach, and so complete the remaining hundred yards to the mountain. A spur affording cover from fire, a lot of us made for it and found the Scots Guards lying there in a huge mass waiting for the other Guards to secure their points of attack further to the right. Their pipers played the *Cock of the North* but as they were lying flat on their backs the music was also rather flat. Finding the Guards blocking our advance here our Colonel (H. C. Denny) began to extricate his men telling the officers of the 58th to take them on.'

From then on the battle evolved into a number of isolated actions as small groups of infantrymen fought their way up the *kopjes* working from cover to cover, firing independently whenever an enemy showed himself. By ten o'clock the Boers had been driven from the heights and the battle won; truly a different kind of fighting from that of eighteen years before. That the 58th had learnt from experience can be judged by their slight casualties, as Barton noted:

'The Grenadiers seem to have had terrible loss. We had drummed it into the men that they must keep widely extended so that they kept about six or ten paces apart while the Grenadiers were crowded to one pace apart and suffered accordingly.'

At Graspan two days later, the Battalion was split; four companies escorting the guns on the extreme left; three companies supporting the KOYLI, who with the Naval Brigade made the main attack; and the eighth company advancing on the left of the sailors.

Again the wisdom of taking open order was demonstrated, for while the Naval Brigade, who closed up in the attack, suffered heavily, the 58th, advancing with eight yards between men, had few casualties.

In the battles at Modder River and Magersfontein that followed, the 58th were not heavily engaged. But a detachment of A and F Companies under Capt Godley held out for seven hours against a force of 1,000 Boers at Enslin station on 7 December. The musketry training of the Battalion paid dividends as the accuracy of their fire deterred the Boers from rushing the position. Capt Godley received the DSO for this action.

Despite the care that had been taken with training in the 58th, the late Victorian soldier found warfare as practised by the Boers a puzzling experience. Barton, whose diaries are full of interest about the campaign, makes some shrewd observations on the soldier of the period:

'No class or race could equal him in his power of standing firm shoulder to shoulder with comrades against a mob of howling savages, delivering his fire steadily and unmoved, his calmness under shellfire. Tommy never budges an inch and as they come swishing along he only smiles and remarks, "There goes another, Bill", or if ordered to retire he gets up and without a trace of confusion moves to wherever he is directed. But modern warfare is just a bit beyond him, he has neither the intellect of a highly educated man, the instinct of a savage or the self-reliance of the colonial. He is a good fellow but a terribly thick-headed one. To think for himself is not what he is accustomed to.'

His endurance, however, was remarkable. After the relief of Kimberley in February, 1900, the 58th spent virtually the whole of that year on the march in often fruitless pursuit of the elusive Boer commandos, first in the Orange Free State, and then in the south-west Transvaal. Week after week the columns slogged along over the dusty, sunburnt veldt, sometimes averaging twenty-six miles a day. Occasionally they were shelled; now and again there would be a brief skirmish with a Boer rearguard; at night, after the long day's march, there was outpost duty; but mostly it was just marching.

Barton described the monotony of this existence:

'Looking back one can recollect but little of what one has been thinking of all the time for one sees just the grass before the eyes. One walks more in a dream with a jingling rhyme or song running in one's head that is difficult to get rid of. The foot soldier has enough to do to go straight on with his equipment and useless rifle on his back. Tommy has now adopted the easiest mode of walk, knees are never straightened up, this is easier for walking and does not jar the legs, shoulder rounded to a very slight stoop to take the weight of rifle and equipment better. This is another instance of the ineptitude of the so-called "soldierly bearing" to real soldier's work.'

By the end of the year the long marches had achieved their object of breaking up the larger Boer commandos; there now remained the task of securing the territory that had been cleared. To this end, a system of blockhouses, garrisoned by small detachments, was established throughout the countryside. For the remainder of the war the 58th were employed in this static role, distributed over the Transvaal in scattered garrisons, with an occasional foray after small bands of Boers to break the monotony. When peace was signed in May, 1902, the Battalion could look back on two and a half years hard campaigning; they had been spared the disastrous battles in Natal and battle casualties had been light; but their endurance on the march and the men's determination never to fall out gave them an unsurpassed reputation in the Army for this prime infantryman's skill.*

The record of the Regiment's role in the Boer War would not be complete without mention of the part played by the Volunteer and Militia Battalions. In February 1900, a company from the 1st Volunteer Battalion joined the 58th and served with them throughout the campaign. During the first year of the war, the 3rd Militia Battalion sent many drafts to the 58th and on 1 May, 1902, the whole Battalion landed in South Africa in time for a month's duty on the line of communications before peace was signed.

These auxiliary battalions were soon to be more closely linked with the Regulars. As a result of Lord Haldane's re-organization of the auxiliary forces in 1907, the 1st Volunteer Battalion became the

---

* By 1909 the 58th had so perfected their system of march discipline that it was adopted by the whole Army.

4th Territorial Battalion, whilst the 3rd Battalion's title changed from Militia to Special Reserve, with the prime function of training and providing drafts in time of war. Both units were soon to find serious work ahead of them.

CHAPTER 8

The Great War

G ENERAL mobilization in August, 1914, produced a rapid expansion of the Regiment. The 4th Battalion mobilized immediately and later the 2/4th and 3/4th were raised as draft-producing units. The early months of the war also witnessed the massive response to Kitchener's call for volunteers for his New Armies, which resulted in the formation, from scratch, of the Service battalions. During the war nine * new battalions of the Regiment were formed, of which three, the 5th, 6th and 7th, were to see active service.

The formation of the 7th Battalion is typical of the improvization that attended the birth of these units. With a group of 200 footballers collected by Edgar Mobbs, a Northampton Rugby International, Guy Paget, an ex-Regular officer of the Scots Guards, arrived at Shoreham to find 1,000 men and a few tents:

> 'I divided them into four lots according to where they came from. I called for old NCOs and about five stepped out. Then about a score of old soldiers broke rank and they were all made Sergeants on the spot. When I got to the footballers, not a man had moved. I made Edgar Mobbs CSM, his four best friends Platoon Sergeants and all the Boy Scouts were made Corporals. The five old NCOs became RQMS and CSMs. We got a master-cook from my old Regiment and later understood why they had been able to spare him, when I found him pursuing his staff, fighting drunk, with a meat chopper.'

A year later this battalion was in action.

Appropriately the 48th, as the senior battalion of the Regiment, was first in the field. As part of the First Division they took part in the Retreat from Mons, covering 140 miles in thirteen days, followed

* 2/4th, 3/4th, 5th, 6th, 7th, 8th, 9th, and 1st and 2nd Garrison Battalions.

by the advance to the Aisne. To be advancing instead of retreating completely re-animated the men, as Lt Needham noted: 'They were singing, laughing and joking as if it was their first day's march instead of having been steadily on the march since 21 August.' Thirteen days later, after the bitter fighting on the Chemin des Dames, over 300 of them were casualties.

After receiving reinforcements, the 48th moved north and from 21 October were heavily engaged in the first Battle of Ypres which culminated in the repulse of the Prussian Guard on 11 November. The remains of the old pre-war 48th came out of this action commanded by a Special Reserve subaltern and, of the officers who had left England three months before, only the Quartermaster, Lt Hofman, was still with the Battalion.

The outbreak of war had found the 58th in Egypt but, by 5 November they were in France as part of the 24th Brigade of the Eighth Division. On 14 November they went into the line for the first time, as Lt-Col O. K. Parker, then a subaltern, remembers:

'Very dark and raining. No previous recce, no orders but to keep closed up (men in full marching order, officers carrying swords). Single file crossing muddy fields, blundering down willow-lined ditches half filled with water. Continual checks, whispered messages being passed, intense rifle fire on the right, odd bullets whining or smacking, considerable bewilderment.'

After a miserable winter in water-logged trenches, the 58th fought their first major action at Neuve Chapelle in March, 1915. The Battalion attacked at dusk on the 10th and, despite suffering heavy casualties, made some progress until the darkness forced them to dig in. Further attempts to advance were made next day but owing to ineffective artillery support, every attack was decimated before it had gone a few yards. At dawn on 12 March the Germans counter-attacked in massed formation. Opening rapid fire, the 58th stopped the attack within 200 yards of their trenches and promptly attacked again, taking advantage of the German repulse. The Battalion captured the enemy trenches but, it proving impossible to hold them, they withdrew to their former position. In two days and nights, the 58th, who had gone into action with nineteen officers and 594 men, had lost seventeen officers and 414 men, of whom 111 were killed.

In only eight months of fighting both Regular battalions had been virtually wiped out. Built up to strength again, by May they were ready to repeat the process. This time it was to be accomplished even more rapidly – in the space of twelve hours. At dawn on 9 May, at Aubers Ridge, the 48th and 58th, both at full strength, went over the top in their respective sectors. By the end of the day the 48th had lost 558 officers and men and the 58th, 426. In both battalions' sectors the story was much the same; the preliminary bombardment failed to cut the German wire and left the enemy machine-guns unscathed; as the infantry attacked they walked into devastating rifle and machine-gun fire. A sergeant of B Company, 58th, discovered that he was the only man in his platoon still on his feet after advancing a mere seventy yards. In both battalions a few men reached the enemy trenches, but the majority were dead or wounded in 'no-man's-land'; those still alive praying for darkness so that they could get back to their lines. The tragic irony of this disastrous day was that, although all ranks had been full of confidence as to its outcome – Lt-Col Mowatt of the 58th had told his men they would pick out the Germans 'like periwinkles' – the security measures had been so poor that the enemy were forewarned of the offensive.

During the next four months the losses of 9 May were made good and a period of trench warfare followed. From 25–28 September the 48th were involved in the fierce fighting around Loos, again suffering severely, not only as a result of the uncut German wire and deadly machine-gun fire, but also from their own gas which hung about between the British trenches and the objective. In this action Capt A. Moutray Read won the first Regimental VC of the war. Although partially gassed himself, he went out several times to rally groups of men of various units and moved freely about under a withering fire, encouraging the men to return to the firing line. While so doing, he was mortally wounded.

The Battle of Loos saw the baptism of fire of the 7th Battalion. Under the command of Lt-Col A. Parkin, they arrived in France on 1 September, 1915 as part of the Twenty-Fourth Division. It had been decided, as an experiment, to use this Division at Loos without any previous battle experience. It was to prove a costly decision. After a hard march up to the front and only a short rest, the 7th were launched into action for the first time as darkness fell on 25 September. The only orders Col Parkin received were 'Follow this man and hurry up.' For nearly three days, under constant shell-fire and attack, without food or water, short of ammunition and soaked to the skin, the totally inexperienced Battalion tried to hold an exposed flank, losing fifty per cent of its strength, including the Commanding Officer. Only two officers of the Battalion had ever been under fire before and of those that survived unscathed, not one had more than a year's service.

Another Service Battalion, the 6th, had an easier initiation to the war. Arriving in France in July, 1915, as part of the Eighteenth Division, it spent the winter of that year in trenches and in the spring of 1916 moved to the Somme, in time for the great offensive in July. The Battalion's supreme moment came on 14 July when, with the 12th Middlesex, they attacked and captured Trones Wood after savage close-quarter fighting and held it for two days until relieved. The advance on the closely defended wood was materially assisted by the action of Sgt William Boulter, a former draper's assistant. Though wounded in the shoulder, he advanced alone under heavy fire across open ground and destroyed a machine-gun post which had been holding up the attack. For his gallantry he was awarded the VC. Two months later the 6th attacked Thiepval with heavy loss, including that of their Colonel, G. E. Ripley, a man of well over fifty. By the time they reached the objective, the companies were commanded by two second-lieutenants and two sergeants.

The 58th entered the Somme battles on 7 July, when the 24th Brigade attacked Contalmaison. The Battalion again suffered grievously and after three days' fighting was down to four subalterns and just over twenty men. Following this action they moved to a rest area where, for the first time in the war, they met the 48th. The latter were moving up to the area the 58th had just left and went into the line on 18 July. On the 20th they attacked near Pozières in a supposedly surprise attack, but found the Germans ready for them. From 16–21 August the 48th were continually engaged in heavy fighting near High Wood, holding their positions despite casualties of 374. Not until 28 September did the Battalion leave the Battle of the Somme; both they and the 58th returned to spend the winter in trenches full of mud, water and slush.

In the 7th Battalion, Edgar Mobbs – only eighteen months after he had enlisted – was now commanding. He had already been wounded at Loos and, as the Battalion was about to attack Guillemont on 17 August, 1916, he was hit again. His beloved battalion had perforce to go through their share of the Somme fighting without him. He was back in command by October but was wounded once more in the attack on the Messines Ridge in June, 1917. By 31 July he was fit enough to lead his battalion into the third Battle of Ypres. Their advance became held up by a German machine-gun post. Col Mobbs, armed with a bag of grenades and accompanied only by his runner, went forward against the post while another party worked round to a flank. Thirty yards from his objective, a burst of fire caught him in the neck. Within ten minutes he was dead, but before he died, he gave his runner the map reference of the enemy position to take back to headquarters. Thus fell one of the outstanding wartime officers of the Regiment. In a tribute, the Lord Lieutenant of Northamptonshire said: 'The power he showed in imbuing his men with his own gallant spirit is a cherished possession not only of the Battalion, but of the county at large.'

On the day Mobbs fell, another gallant officer of the Regiment met his death. During the 58th's assault on Bellewarde Ridge, in the same battle, Capt T. R. Colyer-Fergusson attacked and captured an enemy trench, accompanied only by Sgt Boulding and five men. With this small party he resisted the German counter-attack; then, together with his orderly, Pte Ellis, he advanced to capture an enemy machine-gun, turning it on the Germans. Later he repeated the operation with his sergeant, only to be shot through the head by a sniper as he consolidated the position. His bravery secured a vital part of the German line and he was awarded a posthumous VC.*

After this successful attack the 58th were relieved, but were back in action again at Westhoek Ridge on 16 August. Here Capt Oldfield

* Boulding and Ellis each received the DCM.

was killed. A lance-corporal in 1914, he had survived the slaughter of Neuve Chapelle and Aubers Ridge, received a commission, won the MC twice within three months, but now, as the Battalion consolidated, was shot through the heart as he went forward to reconnoitre. After Third Ypres the 58th enjoyed a fairly quiet period, ending the year in trenches at Passchendaele.

Meanwhile, the 48th had endured another disaster in an almost isolated action, far from the main battlefields of France and Flanders. On 4 July, 1917 they were holding a position which made them the penultimate battalion on the extreme left flank of the Western Front. Between them and the North Sea lay only the 2/60th Rifles; behind and parallel to them was the Yser Canal; on their right flank was a dyke which flowed into a canal. The positions held by the two battalions were contained within an area half a mile long and 700 yards across and consisted, not of the normal trenches, but of breastworks constructed amongst the sand-dunes. The sector, though isolated, was regarded as a quiet one and the 48th looked forward to a peaceful spell beside the sea. They were unlucky, for this part of the coast was, in the words of Philip Gibbs, the war correspondent, to be 'consecrated by one of the most noble and most tragic episodes in the history of this war.'

Just after dawn on 10 July the German heavy artillery opened a hurricane bombardment on the 48th and 60th. Throughout the morning it increased in intensity, systematically deluging each part of the two battalions' areas and maintaining a steady barrage along the canal, thus preventing any chance of retreat or reinforcement. The breastworks were soon demolished; all communications were cut; small arms were clogged by flying sand; and men sought what cover they could find in the shell holes, waiting desperately for the coming attack. Throughout the afternoon the shelling went on. An eyewitness wrote: 'It seemed as if a shell was falling on every foot of ground. The earth was rocking; the smoke and sand were so dense one could hardly see a yard in front.'

At eight in the evening the long-awaited attack came in. Preceded by flame throwers, the 1st German Marine Division swarmed onto the position in five waves. The wrecked defences were rapidly over-

17. The 48th on manoeuvres near Aldershot, 1892. Note the experimental white helmets issued to certain regiments in 1889 to replace the blue type.

18. The 58th Company of the 4th mounted infantry in the Boer War. Note the mixed headdress, on which the black flash (see Appendix) can just be perceived.

19. Captain T. R. Colyer-Ferguson of the 58th, who won the V.C. at Bellewarde Ridge, July 1917.

20. Lieutenant-Colonel Edgar Mobbs, D.S.O., captain of the Northampton Rugby Club and England International, killed in action while commanding the 7th Battalion, 1917.

21. The 48th on operations in Waziristan in 1936. It is interesting to compare the clothing and equipment with that worn in the Tirah forty years earlier.

run but small groups of survivors fought on from the shell holes. Capt Aylett of C Company had ordered his Lewis gunners to button their tunics around their guns to protect them from the flying sand, and these were the only machine-guns still functioning in the Battalion. By nine o'clock those still alive were forced to surrender.

Capt Hayes, the medical officer, described what he saw as he emerged from his shattered aid post:

> 'The sight was extraordinary; the bank of the canal was boiling up in great spurts of earth ... our trenches and dug-outs had been completely obliterated; B Company's headquarters showed up as a tangle of shattered timber ... the ground was just a pattern of interlacing shell holes.'

Nine men only managed to escape by swimming the canal to join the rear details of the 48th, who were then charged with reforming the Battalion.

During the great German offensive of March 1918 both the 6th and 7th Battalions, as part of Gough's Fifth Army, were involved in the heavy fighting of those desperate days. On 22 March a small party of the 6th Battalion under 2/Lt A. C. Herring attacked a German bridgehead on the banks of the Crozat canal and held the position against continuous attacks for eleven hours. In this, his first action, Herring showed great heroism and leadership; he was subsequently awarded the VC. Shortly afterwards the 6th attacked at Hangard to save several French batteries and received a citation in French Army Orders.

The 58th entered the battle on 23 March when the Eighth Division covered the retreat of the Fifth Army across the Somme. After a week's fighting withdrawal, they received a short rest but, on 24 April, were thrown into the counter-attack on Villers Bretonneux, attached to the 13th Australian Brigade. The operation was brilliantly executed but the Battalion had heavy casualties, including the loss of Lt-Col Latham; he had served continuously with the 58th since May, 1915, having joined as a subaltern, aged forty-three.

Lt-Col C. G. Buckle took over the Battalion and the 58th moved to a quiet sector on the Aisne, an area in marked contrast to the shattered battlefields of the Somme. The British IX Corps, of which the Eighth Division, and therefore the 58th, formed part, was now put under the French Sixth Army. The Eighth Division held a salient on the enemy side of the River Aisne; as the 58th were the centre battalion of the centre brigade of the Division, they held an unenviable position.

At one am on 27 May the Germans opened a massive bombardment, saturating the salient with high explosive and gas. For four hours this continued and when the attack came in, the 58th found themselves completely surrounded, the German infantry having broken through on the flanks behind them. The 58th and 1st Worcesters fought to the end, but having lost so heavily from the gas and shelling they could not prevail. Of the entire Battalion only the Adjutant and RSM escaped. Col Buckle was last seen alive in shirtsleeves, revolver in hand, facing the Germans outside his dug-out.

After reforming once again, the 58th were back in the line in July but, compared to their previous experiences, the rest of the war was, for them, without major incident. October was spent in open fighting, advancing thirty miles in three weeks – a welcome relief after the years of static trench warfare.

The 6th Battalion took part in the final attack on the Hindenberg Line in September, where L/Cpl Albert Lewis won the Battalion's third VC by destroying single-handed two machine-gun posts which were holding up the advance. Sadly, he was killed three days later. In October Capt F. W. Hedges also won the VC, and Sgt Gibson the DCM, by capturing six machine-guns and fourteen prisoners. After continuing the pursuit, the 6th ended the war at Le Cateau; there, demobilization started.

The 7th Battalion began the final advance from the Loos area and saw some heavy fighting, particularly at Cagnocles and Haussy, where A Company counter-attacked at night to avoid being surrounded. The Armistice found them at Tournai, where they were gradually demobilized until the final cadre of the Battalion returned home in June, 1919.

The 48th was the only battalion of the Regiment to enter Germany as part of the Army of Occupation. After the Dunes disaster in 1917, the Battalion had been reformed under Lt-Col G. St. G. Robinson,* who had gone out with the 48th in 1914 as a subaltern, and under his command had fought through the German attacks on the Lys in April, 1918. Having switched to the offensive in the St Quentin area in September, the Battalion was in action throughout the latter half of October and was fighting on the Sambre Canal until six days before the Armistice. The 48th's march to the Rhine began on 16 November, reaching Duisdorf, near Bonn, on 24 December. Shortly after their arrival, Col Robinson marched his Battalion to the Rhine, where he and the Adjutant, Capt McNaught – who, like him, was one of the few survivors of the pre-war 48th, having gone out in 1914 as Transport Sergeant – dipped the Colours in the river.

Before closing the account of the Regiment in the Great War, mention must be made of the two other battalions that saw service. The 5th had been formed in September 1914, as a Pioneer battalion; as such its task was mainly the construction of defences in the forward areas, but its picks and shovels frequently had to be exchanged for rifles and bayonets. Going into action near Ypres in May 1915, it fought on the Somme, at Cambrai in 1917, and in the great battles of 1918, particularly at Epéhy, where, fighting as infantry, they suffered 129 casualties.

The 4th (Territorial) Battalion was the only unit of the Regiment to serve outside Europe. After four months of trench warfare on the Gallipoli Peninsula in 1915, it spent the next year defending the Suez Canal. In April, 1917, as part of the Fifty-Fourth Division, it suffered heavy casualties in the second Battle of Gaza, but in November participated in the third and victorious battle for that place. On 27 November, prompt action by Lt-Col John Brown, in ordering a counter-attack when the Turks attacked at Wilhelma, avoided a serious threat to the Jaffa–Jerusalem road, which supplied the troops operating against Jerusalem. The 4th saw further action in 1918, culminating in the offensive in September, which finally broke the Turks and led to their surrender on 31 October.

In four years of war, 6,040 members of the Regiment lost their

* Later Major-General and Colonel, The Northamptonshire Regiment, 1943–1956.

lives. On the credit side, the shared experience strengthened the hitherto rather tenuous regimental link, ordained in 1881, between the 48th and 58th, and furthermore, through the calls made by war upon the men of Northamptonshire, the association between County and Regiment became, for the first time, significant.

CHAPTER 9

Between the Wars

T HE Great War had interrupted the 58th's tour of foreign
service and so, once the Battalion was again on a Regular
footing, it was despatched in 1919 to India for a return to
'proper soldiering'. This activity, so dear to the hearts of Regular
soldiers, was to be denied to the 48th who, after four years conflict
with the Germans, now faced further conflict with the Irish. For
eighteen months the 48th were involved against the Sinn Feiners in
internal security operations of a type now all too familiar, but which,
in 1920, were a novelty to most British soldiers.

After the evacuation of Southern Ireland in 1922, and a short
spell in England, the 48th began to fulfill the promise of a well-
known recruiting poster of the period – 'Join the Army and see the
World'; or, as a wit in the Regimental Journal wrote, 'Join the 48th
and see it twice!' In 1927, after civil war had broken out in China,
the Battalion was sent at short notice to reinforce the garrison protecting British lives and property in Shanghai. After a year there, the
Battalion moved to Malta but within eleven months found themselves on internal security duties in Palestine, where trouble had
erupted between Jews and Arabs. In 1931 the 48th changed station
again, this time to Moascar on the Suez Canal, where it began to
settle down to a peacetime routine.

A quiet life was not to be theirs for long, for the 48th was about
to achieve a 'first' in British Army history. The movement of troops
by air is now commonplace but in the early thirties it was almost
unheard of. Small bodies of men had been transported from place
to place by air, but for a battalion or larger formation, the customary
conveyance was ship or train. In June, 1932, Lt-Col T. S. Muirhead
was ordered to move the 48th from Egypt to Iraq, then under
British mandate, in order to assume the defence of the RAF airfields
whose usual protectors – a local force under British officers named

the Assyrian Levies – were threatening mutiny. Muirhead was further told that, in view of the urgency of the situation, the move would be carried out by air in nine Vickers 'Victorias' of Nos 70 and 216 Squadrons, RAF.

On 22 June, after ten days of order and counter-order, the first troops took off from Moascar at four-thirty am, each aircraft carrying, on average, eighteen fully-equipped men. A company commander afterwards wrote: 'It was an inspiring sight to see these 'planes disappearing into the sky just as dawn was breaking and it was odd to think that there was a company of one's own battalion up in the air. One is so used to seeing them going away on foot, lorry or train, but never before in the air.' Despite the novelty of such an operation, everything had been catered for, even to the extent of issuing each man with two paper bags from the NAAFI, although as one officer wrote, 'in some cases two were not enough.' By 28 June the whole Battalion was safely in Iraq and the companies deployed at their destinations. One or two aircraft had had to make forced landings but on the whole the move went smoothly and the troops on landing seemed none the worse for their unaccustomed means of transport, thus surprising the RAF, who had anticipated having to remove large numbers by ambulance!

With the timely arrival of the 48th, the situation calmed down, the Levies were disarmed, and by the end of July the Battalion was back in Moascar. This first move of an infantry battalion by air across nearly a thousand miles of mountain and desert was, at the time, a noteworthy achievement and, as the AOC Iraq remarked: 'There is evidence that it created a profound impression abroad. It has been an entirely new departure in military transport.'

At the end of 1932 the 48th arrived in India and four years later moved to the North-West Frontier to join the garrison at Razmak in Waziristan. Here they were to spend a year on operations, directed against the followers of a well-known malcontent styled the Faqir of Ipi, who was destined to figure in Regimental mythology for many years to come; even the six years of global war that were soon to follow failed to obliterate his memory.

The method used to pacify this turbulent countryside was to

move columns through the area, engaging the tribesmen as they appeared and, by constructing roads and improving communications, so open up the terrain as to make movement for the enemy correspondingly more difficult. The secret of success of this type of operation was efficient picquetting, a process described by an officer in the Regimental Journal of the time:

'At five am, a Gunner trumpeter sounds Reveille. Breakfast is ready about five o'clock. There is tea, bread, herrings in tomato, tinned sausages and porridge, too, if you can face it. The herrings in tomato look like a bowl of blood in the half light. Soon the column moves on its way. First go the picquetting troops; half doubling, half shuffling, they push out ahead as rapidly as possible. Every hill or commanding feature from which the Pathans can threaten the progress of the column is garrisoned by anything from a Platoon to a Company, depending on the size of the feature. It is not merely a matter of running up a hill and sitting there, because the hill may be already held, so every picquet going up is covered by machine-guns and mountain guns so that at the slightest sign of opposition fire can be brought down immediately. It is a slow business and every hill from which the enemy could interrupt the passage of the column must be seized and held until the rearguard appears. Some of the hills are, perhaps, 1,000 feet above the track and the going is difficult and precipitous. It may be forty minutes before the picquetting troops reach the top, and the column cannot move forward until they are in position.'

As each picquet was withdrawn by the rearguard commander, it was sent forward to rejoin its unit and was then ready for the next picquet on the line of advance. At the end of the day's march a stone perimeter wall had to be constructed, and even after that the tired troops often faced a disturbed night as enemy snipers fired into the camp.

In the Waziristan operations men of the 48th covered 1,500 miles on their feet; they endured hot sun, heavy rain and blinding snow; they faced a cunning enemy, skilled in the use of ground and accurate with the rifle, and whose expertise frequently placed the troops in grave danger. On 8 September, 1936, a picquet of B Company re-sisted attacks for six hours. Pte Letts – who had been christened Aubers Ridge, presumably in memory of the Regiment's action on

9 May, 1915 – won the DCM this day for gallantry under fire and Ptes Clarke and Lee both gained the MM for maintaining their machine-gun in action throughout the day, though both were wounded. Two other men, Ptes Millard and Rose, also won the MM on an earlier occasion for crossing hostile and unfamiliar country in the dark to obtain orders for a picquet that had been forgotten.

On 30 December, 1937, the 48th left Razmak for Dinapore, where they were to remain until the outbreak of war in 1939. In a message to the Colonel of the Regiment, the GOC Waziristan said that the Battalion had 'done very well indeed' and that 'the old 48th reputation has been fully maintained'.

The 58th had moved to the Sudan in 1926 and a year later, when the 48th went to the Far East, returned home. The thirties were not an encouraging time for the Regular Army, particularly for the units at home, but despite the difficulties the 58th secured a fine reputation, both at Aldershot and Ballykinler, scoring notable successes in sport and maintaining the Regiment's name for good shooting.

From 1935–1938 the prowess of the 58th's shooting teams improved steadily both in Command Small Arms Meetings and at Bisley. In 1937 and 1938 they were second and third respectively in the Queen Victoria Trophy, an award presented for the best aggregate scores in various matches of the Army Rifle Association Non-Central Championships. In the 'Abroad' series of the same championships, the 48th had done even better, winning the QVT four years in succession, including the 1937 competitions which were shot at Razmak under very tricky wind conditions, and with picquets on the hills to protect the firers from tribesmen.

The inter-war years saw changes in the Territorial battalions of the Regiment. Before the Great War there had only been one, the 4th, which, by 1919, had been reduced by demobilization to a cadre of seven all ranks. In February, 1920, the Battalion was reformed with companies at Northampton, Rushden, Wellingborough and Kettering. At the same time another battalion was formed, the 5th (Huntingdonshire), with headquarters eventually at Peterborough. By 1937 the 5th had become the only Territorial battalion of the Regiment, for in that year the 4th was converted to a searchlight unit of the Royal Engineers, though retaining links with The Northamptonshire Regiment. However, in April, 1939, the Government belatedly decided to increase the Territorial Army and a new 4th Battalion was formed. By strenuous efforts the Battalion was built up and by August, a month before the outbreak of war, it went to camp over 400 strong.

Two other items of Regimental importance of this period remain to be recorded. In May, 1937, HRH The Duchess of Gloucester, whose home lay in Northamptonshire, consented to become the Regiment's first Colonel-in-Chief, an appointment which gave, and has continued to give, lasting pleasure to all Northamptons. Also in the Thirties, formal alliances were established with Regiments in the Dominions. As a result of the 58th's long connection with New Zealand, an alliance had been approved in 1913 with the 15th (North Auckland) Regiment. From 1930 further alliances were formed: with the 48th and 58th Battalions of Australian Infantry; with The Lake Superior Regiment of Canada; and with The Regiment de la Rey of South Africa. The services of the 48th and 58th in all those countries were thus further commemorated.

And so the Regiment moved towards its second great test within twenty years. 'Proper soldiering' was over, never really to return; the closely-knit fellowship of the old Regular Army was, for The Northamptonshire Regiment, gone for ever.

CHAPTER 10

The Second World War

SIX battalions of the Regiment served in the Second World War; two of these, the 50th, later renamed the 6th, and the 70th were solely employed in the United Kingdom base. The Regular and Territorial battalions, however, were all to see active service. The 1st and 4th Battalions were not to meet the enemy for some years, but the 2nd and 5th mobilized immediately and, when the fighting started in earnest in May, 1940, were serving in France with the Fifth and Fourth Divisions respectively.

When the Germans invaded the Low Countries both battalions moved into Belgium. The 5th was the first battalion of the Regiment to encounter the Germans, when, having withdrawn from a position north of Brussels, they occupied a 2,000-yard sector on the River Escaut. After resisting the first enemy assaults on 20 May, A Company on the east bank was withdrawn across the river to join the rest of the Battalion, which came under heavy attack the next day. All German attempts to cross the river on the Battalion's front failed, but in the afternoon the enemy crossed downstream and pressure on the Battalion's left flank increased. Despite serious casualties, including the Commanding Officer, Lt-Col W. E. Green, the 5th held out against increasingly heavy attacks throughout the 22nd until ordered to withdraw in the evening. From 22–30 May the Battalion moved to the coast by way of Ypres, again suffering casualties, so that by the time Dunkirk was reached, the fighting strength was down to two companies. The survivors were embarked and reached Dover on 1 June. By the 23rd the Battalion had reformed and training for the next round began.

As the 5th withdrew from the Escaut, the 58th, as part of a force sent to Arras to check the German drive to the coast, were under attack on the River Scarpe. On the afternoon of 22 May, the battalion area was shelled by the Germans and A Company, on the right, withdrew from their positions, being later replaced by C Company who had previously been in reserve. This was later to have unfortunate results for C Company. The following morning the French troops on the right of the 58th withdrew and, with that flank now open, C Company became vulnerable to increasing enemy pressure. B and D Companies were also under attack, as was Battalion H.Q. where Lt-Col J. W. Hinchcliffe, who was later awarded the DSO, the Adjutant, Capt P. W. P. Green, and RSM Goodall were all wounded. D Company were forced out of their positions but, led by Capt D. J. B. Houchin, who later received the MC, they counter-attacked with success. By the afternoon however, the right was so imperilled that the Brigadier ordered a withdrawal. This was successfully carried out under heavy fire except by C Company, who were unable to get away and were captured. Maj J. R. Wetherall, now command-ing the Battalion, went forward to look for C Company but was hit and was succeeded by his cousin, Maj R. M. G. Wetherall.

Four days later the 58th took up a defensive position at St Eloi on the Ypres-Comines Canal as part of the line protecting the BEF's escape route to Dunkirk. On the afternoon of the 27th all three companies were heavily attacked and during the night A Company were overrun. At first light a counter-attack was attempted but could not make any headway, and the two remaining companies sustained grave casualties. Maj Wetherall was killed and was succeeded by Maj C. J. M. Watts, who tried to contact Brigade HQ. While he was away, Battalion HQ was strongly attacked but fought on until all ammunition was expended. About 150 survivors of the rifle com-panies managed to escape from the now-overrun position and these, together with the other remnants of the Brigade, were formed into a composite unit under Maj Watts for the final withdrawal to Dunkirk.

After reforming and training for twenty months in the United Kingdom, the 58th embarked in March, 1942, for the capture of Madagascar from the Vichy French, to forestall its possible seizure by the Japanese. The 58th did not participate in the initial landings but on 6 May they attacked Antanambao by night. Information about the ground was scanty and there was little time available for reconnaissance but, despite some confused fighting in the dark, all objectives were taken. Maj Houchin received a bar to his MC for gallantry and Capt J. A. Purton was awarded the same decoration.

Leaving Madagascar in June the 58th spent the next year in India, Persia, and Egypt where, after intensive training for assault landings, they joined the Eighth Army in June 1943.

Meanwhile the 5th Battalion, now in the Seventy-Eighth Division, had been fighting the Germans in Tunisia. Landing at Algiers in early November 1942, they took part in the first attempt to break through to Tunis by attacking first, the key town of Medjez-el-Bab, and second, the village of Djedeida, sixteen miles from Tunis. Medjez was taken but the attack on Djedeida, which was led by A and D Companies mounted on American tanks, ran into heavy opposition, and with inadequate artillery support and under continuous air attack the 5th could not prevail. They tried again next day but were forced back by intense fire.

The Battalion remained on the defensive until 23 December when, in torrential rain, they made a diversionary advance through the mountains in support of an operation which, unknown to them, had been cancelled. They spent a very cold and wet Christmas, short of rations, completely out of touch with Brigade HQ, but beating off all German attempts to dislodge them, and eventually extricating themselves from their dangerously exposed position.

After withstanding German attacks in the Goubellat area from January–March, the 5th returned to the offensive in early April with an advance through the mountains to clear the Medjez road. In the week's fighting that followed, the Battalion made five attacks over very steep, rocky ground, thickly covered in scrub, where all supplies and support weapons had to be mule-borne. Casualties were heavy, particularly in the assault on Tanngoucha, where A Company under Capt C. A. Emery, after attacking at night in thick fog, found themselves and a company of the Lancashire Fusiliers at dawn in full view, and almost in the centre, of the German positions. They were sniped and mortared and finally suffered an intense bombardment from their own artillery, under cover of which the Germans rushed their positions. As Emery later wrote: 'We were given a demonstration of the axiom that in the infantry nothing is ever so bad that it couldn't be worse.'

The 5th's final task in the campaign was to capture Sidi Ahmed, just north of Longstop Hill. The companies had to advance across open ground without tank or artillery support and soon were held up by heavy artillery and mortar fire. They attacked again next day and eventually achieved their objectives, but had to withdraw to avoid being surrounded by a strong counter-attack. Like the rest of the Seventy-Eighth Division, which had borne the brunt of the campaign, the 5th did not take part in the final assault on Tunis, although they had the satisfaction of occupying the city the day after it was entered by the Seventh Armoured Division. Thus ended an arduous six months' fighting which had cost the Battalion 740 casualties.

The 58th and 5th Battalion now found themselves both in the Eighth Army for the invasion of Sicily. This was to be the 58th's first encounter with the Germans since 1940; furthermore the close country in which they were to fight was very different from the deserts in which they had been training. Now commanded by Lt-Col J. A. W. Ballard, they landed on 10 July and, after brushing aside some Italians, they entered Syracuse the same evening. Next day, however, they came up against the Hermann Goering Division amongst the orchards and olive groves in front of Priolo. Some very stiff fighting ensued as each company was thrown into the battle but, although the Battalion got within a few hundred yards of the village, no progress could be made across the final stretch of fire-swept open ground. A further attack was planned for the following day but during the night the Germans withdrew.

Over the next two days the 58th cleared the area around Augusta, taking 350 Italian prisoners. The advance from the coast had meant a hard slog for the rifle companies, for the heavily-laden troops had marched and fought in heat and dust over rocky ground in boots softened by immersion in sea water. Subsequently the Battalion held an area on the River Simeto, as part of the holding role then allotted to the Fifth Division and finally, when the advance was resumed, ended their share of the fighting by occupying Misterbianco.

While the Fifth Division held its ground, the Seventy-Eighth Division, hitherto in reserve, was brought up to attack Centuripe, the key to the German position south and west of Mount Etna. The 5th Battalion, having landed on 25 July had already made one attack in support of the Canadian Division. On 1 August the Battalion led the advance of the Seventy-Eighth Division on Centuripe and the next day captured a hill in rear of the town. The position had been extremely strong with steep, terraced hillsides, covered in vineyards and scrub which demanded great stamina and alertness from the assaulting infantry. Advancing again on the 8th, the Battalion had some stiff fighting in Bronte, where Lt-Col Buchanan was severely wounded, followed by a further advance until 14 August, when the Seventy-Eighth Division's task in Sicily was completed.

After landing in the toe of Italy, the 58th experienced some hard fighting on the Sangro in the winter of 1943 and in January moved to the Fifth Army front on the west coast for the crossing of the Garigliano river. The 58th's task was to capture Minturno after a bridgehead had been secured by the 6th Seaforth. However, when the Battalion advanced, they found that the Seaforth, in the face of stiff resistance and extensive minefields, had only gained a precarious footing on the far bank, and, instead of advancing to their objective, the 58th had to fight to widen the bridgehead. Thirteen days later they attacked and held Point 156 on the Minturno ridge, a key position which had hitherto defied all attempts to take it. As a result of these operations, three MCs, one DCM, and one MM were awarded to members of the Battalion.*

From 9 March to the end of May the 58th served in the Anzio beachhead. This was the hardest fighting the Battalion endured in Italy, and none was harder than the nine days spent in the 'Fortress', the most vital sector of the beachhead; an undulating area of scrubland, much of it charred and dead, seamed by deep vertical-sided wadis, covered in litter and pervaded by an overpowering smell of decay. In places the forward slit-trenches were only thirty yards from the Germans, being under almost constant fire from mortars,

* Capts E. W. Kitchen, A. C. Garner, Lt S. C. Hamer; Sgt V. Bell; Cpl J. Manners.

rifle-grenades and snipers. On 30 April the forward platoon was overrun and all counter-attacks failed to retake it despite a gallant attempt by A Company, for which Sgts Heward, Organ and Underwood were all later awarded the DCM.

Three weeks later A Company fought another brisk action on the Moletta river, during which No. 9 Platoon, commanded by Sgt Bell, who had previously won the DCM on the Garigliano, engaged a superior force in hand-to-hand fighting to cover the Company's withdrawal. The gallantry displayed in this action was recognized by the award of the MC to Maj J. C. Denny, and the MM to Sgt Bell and Cpls Wilford and Doubleday. This was the last important action of the 58th in Italy for, after reaching Rome, they were withdrawn for a rest in the Middle East.

The 5th Battalion started the Italian campaign at Taranto and ended it on the River Po. On a chill November night in 1943 they waded across the icy, racing waters of the Sangro to attack and hold positions in incessant rain and bitter cold. At Cassino in March they defended a narrow, jagged ridge, with forward positions only ninety yards from the Germans and the rear dominated by enemy observation posts. In May, still at Cassino, they had to fight hard to secure their own start line before joining in the main attack. June saw them in a savage house-to-house struggle in the narrow streets of the hilltop town of Montegabbione, followed by hard fighting under heavy shelling and mortaring around Lake Trasimene.

After a rest in the Middle East, the 5th returned for what was to be one of their most arduous ordeals of the campaign; the fighting round Monte la Pieve and Monte Spaduro, an important mountain feature on the flank of the American drive on Bologna. Between 13 and 15 October the Battalion made three attacks on Pieve; although two platoons got to the summit at the second attempt, the opposition was so fierce and the companies so weakened by casualties, that all efforts were in vain. After the capture of Spaduro, the sector was held for ten weeks throughout the hard winter. An officer of the 5th, Capt Giblett, described the conditions:

> 'Slit trenches were full to the brim with muddy water which had to be baled out, while shelling and mortaring were so frequent it was

sometimes difficult to know whether to risk lying in the open or jump into the trench full of water. Everyone was covered from head to toe in mud, all were thoroughly soaked to the skin and the cold freezing nights caused the utmost discomfort. During the hours of darkness everyone had to make the best of bad conditions for positions were manned 100 per cent. It was not uncommon to find the mechanism of light machine-guns had frozen stiff and much care was given to keeping the guns ready for immediate use.'

When the offensive was resumed in April, the Battalion fought their last battle of the war at Argenta, forcing their way into the town supported by tanks and flame-throwers. Continuing the advance northwards, the 5th met only scattered resistance, sustaining its final casualties eight days before the Germans surrendered.

Meanwhile the 58th, after returning to Italy in March 1945, moved straight on to yet another theatre – Germany. But here the end was in sight, and after some minor operations on the Elbe, the Battalion advanced to the Baltic, ending the war at Lubeck. They were the second battalion of the Regiment to serve in North-West Europe, for in February the 4th Battalion, which had spent the war in the United Kingdom, had arrived in Holland. They took over a sector on the River Maas but towards the end of March were split up to provide detachments for duties connected with the Rhine crossing. Thereafter the 4th remained on the Rhine until VE-Day, carrying out many unexciting but essential tasks.

It is now time to discover how the war had affected the 48th. After four years in India on internal security duties and training for jungle warfare, the call to battle came in December 1943, and under the command of Lt-Col D. E. Taunton the 48th left for the Burma front as part of the Twentieth Indian Division.

Maj E. P. Kelly, then a young soldier who was to win the DCM for his fine work as a sniper, has written of this time: 'The prospects of action after months of training pleased the 48th. We were young, fit and spoiling for a fight. No-one doubted the CO knew his stuff and provided the rations arrived on time there would be few complaints. Life was good and morale was high.' The months of training were to be tested at Kyaukchaw, a heavily bunkered and

22. *The 58th on manoeuvres near Aldershot, 1935. Note the neck curtains fastened to the service dress caps.*

23. *Lieutenant-Colonel A. A. Crook leading a mounted patrol of the 5th Battalion in North Africa, 1943.*

24. Lieutenant-Colonel D. E. Taunton, who commanded the 48th in Burma, giving orders over the radio.

25. The last parade of the 48th and 58th Colours in the Market Square at Northampton in 1960. The 58th Colours are on the left. The Colour Parties are wearing the East Anglian Brigade cap badge.

excellently camouflaged Japanese position, situated on top of the vertical bank of a deep, fast-flowing river.

The fierce fighting that followed was characterized by great determination and gallantry on both sides, especially by Lt A. G. Horwood, who was mortally wounded while standing in the Japanese wire under point-blank fire directing an attack. His outstanding bravery from the first assault until his death two days later was recognized by the posthumous award of the VC. Every possible approach to the position was tried by the 48th but the Japanese field of fire, conveniently cleared by the preliminary air strike, proved impassable. Nevertheless the Battalion held on, harassing the enemy, who eventually withdrew.

E. P. Kelly's memories of Kyaukchaw recall the 48th's first action of the war:

'A march of some sixty miles across bamboo covered mountains and through teak forests to cover a mere twenty-five miles "as the crow flies" was very severe on men and mules alike. It was at Kyaukchaw that we discovered that although the Jap was no longer as "invincible" as we had feared, he was a tough, stubborn fighter who would indeed rather die than surrender. We suffered eighty odd casualties on that first day and the screams of the wounded as they were carried off the hill were not easily forgotten. Stretchers were at a premium. The really badly wounded did not survive the tortuous route back to base. Even so morale remained high. It was the dry season and the 48th had triumphed.'

From March–July, 1944, the 48th were fighting continuously in the battle for Imphal. After providing the rearguard at Moreh, the Battalion moved to Bishenpur where their brigade was to spend three months withstanding the best attempts of the crack Thirty-Third Japanese Division to open the way through the right flank of the defences at Imphal. The Silchar track and the features on either side were the scene of constant attack and counter-attack, ambushes and sniping. The men were soaked by torrential rain and lived in almost perpetual mist. The latter often served to conceal movements from the enemy but had an unfortunate knack of disappearing when it was most needed, as Col Taunton noted in a letter describing the approach to a strong Japanese position:

'It rained extremely heavily throughout the night and our progress was a nightmare. What should have taken three hours with two hours rest took five hours with no rest. The leading company just had time to down their rum and in they went. They took the first objective and hung on to it. The Jap had let a lot of stuff fly around, red tracer, heavy mortars, etc. We got held up by accurate fire on a bright sunny day when we relied on mist and cloud to help us.'

When the 48th were finally relieved after six months' almost continuous fighting, they had suffered nearly 450 casualties, many had dysentery and the Battalion badly needed a rest, as Kelly remembers:

'Fighting in the monsoon season was a swine. In daily contact with the Jap; fighting and existing in unremitting rain; a monotonous diet and an acute lack of sleep. Faces and bodies became thin and drawn; eyes dark-rimmed and sunken – old men's eyes craving for a good sleep. We desperately needed a good hot meal and to sleep the clock round. But morale remained high. We were hardened fighters and even while lying chafing in our sodden clothes we could still dream of that next leave in Calcutta. Yes, life was good – for the living.'

The 48th returned to the front in November to find that the Fourteenth Army had assumed the offensive. After a 250-mile march into Burma, the Battalion came out into open country and in a well-fought action captured Budalin on 10 January. Here No. 2 Company particularly distinguished itself, Capt Cherrington being awarded the DSO for his gallantry and leadership; a decoration which he sadly never lived to wear, being killed ten days later at the capture of Monywa.

The next task of the 48th was to seize a subsidiary bridgehead across the 600-yard wide Irrawaddy to prevent enemy reinforcements moving up to attack the main Twentieth Division bridgehead. Despite the strong current and defective assault craft, the Battalion established a foothold, where they were reinforced by the 9th/14th Punjabis. For three days and nights the positions were subjected to heavy shelling and determined attacks, particularly against No. 4 Company; but fighting desperately, often at hand-to-hand, and without rest, the 48th and Punjabis held their ground. Thereafter no further attacks were made, but it was equally impossible to break out. Cooped up in a small sandy perimeter amongst the elephant grass, under constant shell-fire, the 48th held the bridgehead for three weeks until relieved. Col Taunton wrote of this period:

'I remember the sand, one shell near your trench and you had to dig another, the heat, boredom, constant shelling and attacks were three weeks of battle in its grimmest form. It took four days to get the tired, strained look out of the men's eyes.'

Just before the 48th advanced again, they lost Col Taunton on promotion to Brigadier. He had trained the Battalion in India and led it throughout its campaign in Burma, and the award of the DSO and bar recognized the worth of his leadership. Then, under the command of Lt-Col P. de C. Jones, the Battalion fought its last actions of the war, clearing the road to Kyaukpadaung to finally link up with the Seventh Indian Division near Mount Popa. Thus, after nearly 200 years since it first stood firm in battle against the Jacobites at Falkirk, the 48th fired its weapons for the last time in war against the beaten Japanese.

CHAPTER 11

'Days to Do'

PEACETIME soldiering after the war was quite different from anything ever experienced before. To begin with the Regiment was much smaller. The two Territorial battalions were disbanded in 1946, although the 5th was re-formed two years later. Then, in 1947, as part of the reorganization designed to reduce Regiments of the Line to only one Regular battalion, the 48th, after 206 years of unbroken service, was placed in 'suspended animation', a euphemism of the time for disbandment. A year later, in an effort to disguise this unpalatable decision, a War Office sleight of hand permitted the amalgamation of the two Regular battalions; the 58th thus also disappeared and the resultant hybrid emerged as '1st Battalion The Northamptonshire Regiment (48th/58th)'.

The stations initially occupied by the Regular Battalion were very different from those of pre-war days. Instead of the time-honoured 'outposts of Empire', they found themselves in the exotic European cities of Berlin, Vienna and Trieste; instead of the excitements of the North-West Frontier, there were mock battles on the training areas of Sennelager with the British Army of the Rhine.

There were differences in the Battalion's appearance too. Full dress had never returned after the Great War but even the smart peaked cap and service dress of the inter-war years now gave way to the shapeless beret and 'battle dress', a uniform not particularly well-suited to battle and inelegant on parade.

A pre-war Regular battalion had a homogeneity that now seemed lacking. An urge to matrimony permeated all ranks. The many problems arising therefrom came to occupy an unwarrantably high proportion of company and platoon commanders' time and the call from the married quarters often proved more compelling than that of the martial bugle.

Perhaps the Sergeants' Mess remained much as it had always been, but the officers of the 48th/58th were now very varied. There were pre-war Regulars, the veterans of Razmak and Ballykinler, many of them anxious, with some justification, to return to more familiar soldiering; there were temporary wartime officers who elected to remain in the Service, some of them refugees from the Indian Army, whose presence in The Northamptonshire Regiment was often the result of an arbitrary decision by a War Office official rather than an expression of their own wishes; there were products of the new Sandhurst, the heralds of the coming professionalism, so fashionable in the sixties and seventies; finally, and most varied of all, there were National Service officers who came in all shapes and sizes, a microcosm of middle-class English youth; ranging from an Old Harrovian to one who was overheard castigating the arms drill of a soldier with the admonition, 'Dig that crazy slope'. Despite their unmilitary demeanour, many achieved, in the short time each served with the Battalion, considerable military competence; there were some, however, whose existence as officers was a source of wonder.

The lack of homogeneity was also found in the rank and file, the great majority of whom were conscripts. Herein lay the greatest difference of all, for never before in the history of the Army, had the mass of its troops in peacetime been pressed men – men who in the main were understandably reluctant to serve. These were the years of the National Serviceman, whose battle-cry was 'Days to Do', and whose standard was the 'demob chart'. The few Regular soldiers among the rank and file – and in the Regiment they were not numerous – had to be brave spirits to make a virtue out of soldiering, to show loyalty to their regiment and to the Army, in the face of the all-pervading reluctance around them. Anyone who can compare service in a predominantly conscript unit with that in an all-Regular battalion knows the totally different atmosphere that exists between the two. This is not to say that the National Servicemen were incompetent soldiers; most gave excellent service and did their job to the best of their ability. But, as one military writer has aptly put it, 'it was just the overwhelming atmosphere of resigned distaste, which could often be summoned up even when a man was obviously enjoying himself.' * This it was that prevented a resumption of the family feeling that had prevailed in the pre-war Regiment.

It is a fact of military life that the best means of consolidating a unit is a spell of active service, or at least a worthwhile task that justifies the months of training and administration. This never really came the way of the 48th/58th in the post-war years. In this they were unlucky, for the standard of training and discipline was such that, had the chance presented itself, the Battalion would have acquitted itself as well as did some other predominantly National Service units, who found fame in the many post-war conflicts. They were fortunate to serve in some extremely pleasant stations, but the tasks for which they were trained, by some admirable commanding officers, somehow eluded them. There were riots in Trieste but the Battalion had by then moved to Germany; the war in Korea was over by the time they reached that country; they were called out to face rioters in Hong Kong but the trouble was short-lived – the imposing six feet seven inches and eighteen stone of the officer commanding A Company may have done much to curb the hysteria of the diminutive Chinese; in 1957, while in England, the Battalion was suddenly placed at short notice to move to a secret destination, but the final order to move never came; a year in Aden recalled echoes of Razmak with picquets on the hills and hostile tribesmen on the prowl, but the fighting of the year before did not recur and the worst trouble was yet to come. The 48th/58th had to console themselves with the thought expressed by the Commander HQ Aden Sub-Area, who wrote to the Commanding Officer, Lt-Col G. V. Martin: 'I believe that the behaviour of your men and that their attitude to the local population has been a major factor in preventing serious trouble.'

If the Battalion was not called upon to fire its weapons in anger, it at least showed it had not lost its skill at shooting. From 1947–1956 they won the Queen Victoria Trophy five more times, and were second three times; they won the Machine-Gun Cup three times and some individuals put up fine performances at Bisley. But success in competition shooting depended increasingly on a few enthusiasts

* Lt-Col J. M. Baynes, *The Soldier in Modern Society*, Eyre Methuen, 1972, page 75.

and when these left the Battalion the necessary expertise went with them.

And so the years passed; a frustrating time in many ways for the Regulars, but it was the only soldiering there was and they made the best of it. It was as well they did, for these were the last years the Regiment was to know as a living entity. In 1957 they received the first hints that their future existence might be in jeopardy – the news of the abolition of conscription and the consequent reduction of the Infantry by amalgamations.

The first nail in the Regiment's coffin had been the linking of regiments for administrative purposes into territorial groups in 1947. As a result, the Northamptons found themselves, by a geographical aberration perpetrated by the War Office, in the East Anglian Group, later Brigade. Officers and men could be called upon to serve in any regiment of the Group; this was not popular, but at least the Regiment's identity was preserved. By 1957 the Regiment, like its National Servicemen, had only 'days to do'.

The prospect of amalgamation, when the Regiment's poor regular recruiting was taken into account, was regarded with resignation but not, when contemplating the likely partners, with relish. It therefore came as a surprise, and indeed almost a relief, when the partner was named as The Royal Lincolnshire Regiment – an alternative that had never been considered.

Apart from a similar Regimental March, the two Regiments had little in common. There were thus many problems to be solved by the Colonels of the Regiments * and their Committees, and many interests to be reconciled, not only between, but also within each Regiment, for the views of the retired members were given equal weight to those of serving members, and the two groups were frequently at odds. Discussion over the title of the new Regiment waxed fiercest of all. The retired group were keen to retain the county titles but a straightforward union of the two was thought to be too cumbersome. Some factions wished to resurrect the old numbers, while others preferred to link the title to the name of the Duchess of

* The Northamptons' Colonel was Brigadier John Lingham.

Gloucester, who was to bestow her Colonelcy-in-Chief on the new Regiment. In the end the problem was brusquely settled by the War Office, whose unimaginative choice – which pleased no-one – added to a compromise secondary title, resulted in the new Regiment starting life with the longest title in the Army List: The 2nd East Anglian Regiment (Duchess of Gloucester's Own Royal Lincolnshire and Northamptonshire).*

The first loss of identity occurred in 1958, when the old Regimental cap badge, fondly known as the 'soup plate', was replaced by the East Anglian Brigade badge, on which the Gibraltar castle was still retained, superimposed on the Garter star. The old badge was, however, kept by the Territorial battalion, which was not affected by amalgamation.

In May 1960, after arriving home from Aden, the Battalion paid its last visit to Northamptonshire and the old Colours of the 48th and those of the 58th, which had been carried under fire at Laing's Nek seventy-nine years before, were paraded for the last time through the County. On 1 June the 48th/58th joined the 10th Foot, and The Northamptonshire Regiment ceased to exist as a Regular unit of the Army. The name was to be carried on for a few more years by the Territorial battalion, but in 1969 this was reduced to a cadre, which survived until 1971.†

In an address to the new Battalion, Lt-Col Martin spoke of traditions which 'have been won in peace and in war; against disease and disaster; under conditions that demand the best of men'. The traditions and honours of the 48th and 58th were now joined to those of another. The Regiment had died but its history remained. It has been the purpose of this book to try to ensure it is not forgotten.

---

* Its nearest rival – The Royal Highland Fusiliers (Princess Margaret's Own Glasgow and Ayrshire Regiment) – had thirteen letters less!

† In 1961 the 5th Battalion was designated 4th/5th, following the disbandment of a T.A. Light Anti-Aircraft Regiment, part of which had been the old 4th Northamptons up to 1937. In 1967 the 4th/5th became The Northamptonshire Regiment (Territorial), following the reorganization of the Territorial Army into the Territorial and Army Volunteer Reserve. In 1971 the cadre provided the nucleus of the 7th (Volunteer) Bn The Royal Anglian Regiment.

Postscript

In the early 'sixties it became Army Council policy to encourage the various territorial Brigades of Infantry to convert themselves into 'Large Regiments'. As a result, in 1964 the 2nd East Anglian Regiment became 2nd Battalion The Royal Anglian Regiment, the traditions, honours and customs of the former Royal Lincolnshire and Northamptonshire Regiments reposing solely with that Battalion. In 1969 the four battalions of The Royal Anglian Regiment became more closely integrated, so that now it is that Regiment as a whole, not merely the 2nd Battalion, that is the lineal successor to the 48th and 58th Regiments of Foot.

APPENDIX

A Regimental Miscellany

Regimental Marches

The earliest known march was a tune called '*Wilkes' Release*' or '*Quick March 48th Regiment*' composed for drums and fifes around 1800.

The later march used by the 48th up until 1948 was '*The Northamptonshire*', or more popularly, '*Hard Up*'. It was composed by Bandmaster W. Allen of the Northamptonshire Militia, probably between 1854–1856, when the Militia was embodied during the Crimean War. It is not known when it was adopted by the 48th but possibly not until 1881.

The 58th's march was '*The Lincolnshire Poacher*', though when it was first taken into use is also not known.

REGIMENTAL MARCH PAST
THE NORTHAMPTONSHIRE REGIMENT
(48th/58th)

After 1948 a new quick march combining *'Hard Up'* and *'The Lincolnshire Poacher'* was arranged by Bandmaster S. W. Ord-Hume for use by the Regiment.

Nicknames

Prior to the Great War the 48th sometimes referred to themselves as 'The Cobblers', an allusion to the chief industry of Northampton, but the 'Four and Eights' was more common.

The nickname of 'Steelbacks', acquired by the 58th allegedly as a tribute to their stoicism under the lash, used, in days gone by, to be fairly well-known throughout the Army. The tradition is held to date at least from the Peninsular War, possibly from the Siege of Gibraltar, but in accounts of the Regiment in New Zealand in the 1840s by men of the 58th, the name of 'Black Cuffs' is more common. 'Steelbacks', however, was back in favour at the time of the Zulu and Transvaal Wars and remained in fairly common usage until recent times.

Badges, facings and colours

Until 1881 the 48th had no special device other than its number. The 58th wore the Castle and Key of Gibraltar and the Sphinx with 'Egypt' on various appointments. When collar badges were introduced in 1874, the 48th wore a brass 'Talavera' laurel wreath, and the 58th a brass castle.

In 1881 the Castle and Key was adopted as the main Regimental motif, together with the words 'Gibraltar' and 'Talavera'. The collar badge featured the Cross of St George from the badge of the Northamptonshire Militia and also a horseshoe, the badge of the old Rutland Militia. A crown appeared both on the buttons and the collar badge.

The buff facings of the 48th and the black of the 58th remained in use until 1881. Then, in common with other non-Royal English regiments of the Line, the Regiment adopted white. In 1927 buff facings were restored and the 58th's black was preserved in the officers' waistcoat in mess dress. These two colours, together with pale blue, formed the Regimental colours. The blue's origin is a little uncertain but is supposed to commemorate the voyage from Louisburg to Quebec in 1759 when the 48th served as marines.

A black flash was traditionally worn on tropical helmets since the turn of the century and subsequently on the sleeve of the battle dress blouse. In recent years this was usually held to be in memory of General Wolfe, but its origin was in fact far more mundane, being merely a Regimental distinguishing mark worn on their khaki helmets by the 58th during the Boer War.

Militia and Volunteers

The 18th and 19th Century Militia of Northamptonshire and Rutlandshire can be traced back through the 'train-bands' of Tudor England and the shire levies of the Middle Ages to the Anglo-Saxon *Fyrd*. The origins are thus far more ancient than those of the 48th and 58th. Both Militia regiments were embodied during the Napoleonic Wars and the Northamptonshire Militia garrisoned Gibraltar during the Crimean War. The forces of the two counties were amalgamated into a single battalion of the 'Northamptonshire and Rutland Militia' in 1860, a second battalion being formed in 1872.

There were units of Volunteers and Fencibles in Northamptonshire during the 18th Century and the Napoleonic Wars. But the origins of the Territorial battalions of the Regiment lie in the great Volunteer Movement of 1859 and the formation of companies of Rifle Volunteers in the County; in 1860 these were grouped together to form the 1st Administrative Battalion Northamptonshire Volunteer Rifles. The uniform was grey with scarlet facings, this being retained after converting to the 1st Volunteer Battalion The Northamptonshire Regiment, until the formation of the Territorial force in 1908 when the scarlet uniform was adopted.

Battle Honours

The number of Battle-Honours awarded in over 200 years service are:

	48th Regt	58th Regt	Northamptonshire Regt
18th Century Campaigns	4	4	–
Napoleonic Wars	11	8	–
Victorian Epoch	1	2	3
Great War	–	–	76
Second World War	–	–	32
Totals	16	14	111

Grand Total: 141